FIRE SUPPRESSION PRACTICES AND PROCEDURES

FIRE SUPPRESSION PRACTICES AND PROCEDURES

Eugene Mahoney

BRADY
A Prentice Hall Division
Englewood Cliffs, New Jersey 07632

Library of Congress Cataloging-in-Publication Data

Mahoney, Gene
Figure suppression practices and procedures / Eugene Mahoney.
p. cm.
Includes index.
ISBN 0-89303-215-8
1. Fire extinction. I. Title.
TH9151.M25 1992
628.9′25--dc20 91-13727
CIP

Editorial/production supervision and Interior Design: Lynda Griffiths
Cover Design: Wanda Lubelska Design
Prepress Buyer: Ilene Levy
Manufacturing Buyer: Ed O'Dougherty
Acquisitions Editor: Natalie Anderson

Printed in the United States of America

10 9 8 7 6 5 4 3 2

ISBN 0-89303-215-8

PRENTICE-HALL INTERNATIONAL (UK) LIMITED, *London*
PRENTICE-HALL OF AUSTRALIA PTY. LIMITED, *Sydney*
PRENTICE-HALL CANADA INC., *Toronto*
PRENTICE-HALL HISPANOAMERICANA, S.A., *Mexico*
PRENTICE-HALL OF INDIA PRIVATE LIMITED, *New Delhi*
PRENTICE-HALL OF JAPAN, INC., *Tokyo*
SIMON & SCHUSTER ASIA PTE. LTD., *Singapore*
EDITORA PRENTICE-HALL DO BRASIL, LTDA., *Rio de Janeiro*

This probably will be the last textbook I will write. It is therefore only fitting and proper that I dedicate it to those who have been closest to me during my fire service career—my family. Without their help and encouragement, nothing I have done would have been possible. It is with pride that I acknowledge my family—my wife, Ethel; my son, Mike, my daughter, Jeanne Juncal; my stepdaughters, Lana Hansen, Cathi Glassman, and Jeri Stapp; and my grandchildren, Kelly Evans, Tony Evans, Leah Mahoney, Amber Mahoney, Joshua Juncal, Jeremiah Juncal, Jennica Juncal, Josie Juncal, Ian Glassman, Raychel Glassman, and Nathan Glassman.

Contents

Foreword

Gene Mahoney was released to inactive duty as a pilot from the U.S. Navy in 1946. He served an additional 18 years in the reserve, retiring as a Lieutenant Commander. During his time with the Navy, he flew both reciprocating engine and jet aircraft.

Gene joined the Los Angeles Fire Department as a firefighter in 1947. He retired as a battalion chief in 1969. During his time with the department, he was assigned to various areas of the city, including five years in the downtown area, five years in the harbor area, and five years in the south-central area of the city. As a battalion chief, he served five years in the most active firefighting battalion in the city, additional time in the high-rise area of the city, and as commander in charge of the firefighting forces at the Los Angeles International Airport. His special duty assignments included several years in the training section. At the time of his retirement, he was responsible for the Public Relations Section of the department.

Gene Mahoney retired from the Los Angeles Fire Department to accept the position of fire chief for the City of Garden Grove, California. He was later advanced to the position of Public Safety Director and then accepted the assignment as Assistant City Manager for Public Safety. In these positions, he was responsible for the operation of both the fire and police departments. He left the city of Garden Grove to accept the position of fire chief for the Arcadia (California) Fire Department. He retired from this position in 1975.

Gene was responsible for the development of the fire science curriculum at Harbor College and served there as a part-time instructor for 12 years. He also taught at Long Beach State College for two years. Upon retiring as fire chief from the city of Arcadia, he accepted the position of fire science coordinator at Rio Hondo College. While there, he developed the fire science curriculum into one of the most complete programs in the United States. He retired from Rio Hondo College as a Professor of Fire Science in 1988.

While with the Los Angeles Fire Department, Gene attended the University of Southern California, where he received his BS degree in Public Administration with a minor in Fire Administration, and later his MS degree in Education.

In addition to authoring several articles in professional magazines, Gene has had published several textbooks and several study guides in the field of fire science. Some of the textbooks include *Fire Department Hydraulics, Introduction to Fire Apparatus and Equipment,* and *Fire Department Oral Interviews: Practices and Procedures*. The study guides include one for his text, *Introduction to Fire Apparatus and Equipment;* one entitled *Firefighters Promotion Examinations,* and one on *Effective Supervisory Practices*. He also had a novel published, entitled *An Anatomy of an Arsonist*.

During his career, Gene has been very active in professional and service organizations. He served as:

District Chairman, Boy Scouts of America
President, United Way
District Chairman, Salvation Army
President, International Association of Toastmasters
President, Rio Hondo College Faculty Association

Chief Robert Wilson
Santa Fe Springs Fire Department
Santa Fe Springs, California

Preface

I didn't want to be a firefighter. My father had been one, and my childhood memories of activities around the fire station were not pleasant. It was my mother who encouraged me to take the test. She was more aware of the amenities of the job—particularly the financial security and the work schedule.

I changed my opinion of the job the first day in the training tower. I was impressed with the organization, the professionalism of the officers, and the fun of the work itself. The assignment to a fire station with its clubhouse atmosphere, the comradeship, the excitement of the bell, and the challenge of emergency activities added to convince me that this was where I wanted to spend the rest of my working days.

I started studying almost from the day I entered the service. I read every book on the fire service I could find, and took every class available. By the time I had five years on the job, I thought I knew all the answers. Some 35 years later, when I hung up my turnout coat and put away my boots, I realized how little I really knew. About the only things I could be certain of were:

1. Nothing is constant but change.
2. No two fire departments do everything exactly the same.
3. There is usually more than one way to extinguish a fire or complete a function effectively.
4. Learning is a lifelong experience.

One of the things that should be done before writing a book is to select your audience and establish your objectives. Because of the listed factors, I found this to be extremely difficult. About the only objective I could arrive at was to pick up the reader wherever he or she might be in the learning cycle of firefighting and carry him or her as far as possible in a few hundred pages of the written word. Consequently, I feel that the results are that this book can be used effectively as:

1. A text for an organized course at a community college

2. A guide for the development of training sessions in the fire station
3. A study guide for promotion
4. A reference manual for fire department officers

I feel that the questions at the end of each chapter will be particularly valuable for points 1, 2, and 3.

I wish you good luck, regardless of your use of this book. I hope you enjoy reading it as much as I did doing the research and putting it together.

ACKNOWLEDGMENTS

It is impossible to acknowledge all those individuals who have contributed to this book. There is no doubt that many of the officers and firefighters that I have worked with over the years have unknowingly contributed to various portions of these chapters. The instructors of courses I have taken, and the students of classes I have taught have definitely had an influence on the book's contents. The authors of books I have researched, such as Emanuel Fried, James Casey, William Clark, Charles Walsh, Leonard Marks, and Lloyd Layman, have all provided bits of information, as have the contributing authors of training manuals and magazines published by such organizations as the National Fire Protection Association, the International Fire Service Training Association, Fire Engineering, and the Los Angeles Fire Department. To each of those who have unknowingly contributed, I offer my sincere appreciation.

There are some people, however, who made a substantial contribution in the final steps of preparation. First credit goes to my wife, Ethel, who read each chapter as it was completed and made constructive recommendations for improvement. The efforts and concrete suggestions made by Retired Chief Ron Lathrope, Lynwood Fire Department; Captain Richard Beckman, Senior Drillmaster, Rio Hondo Fire Academy; Chief Gary P. Morris, Tactical Operations, Phoenix Fire Department; Russell J. Strickland, Greenbelt, Maryland; and Jim Linardos, Reno, Nevada, in reviewing the final draft are sincerely appreciated. Gratitude is extended to the following retired members of the Los Angeles Fire Department for sharing their knowledge on ship and waterfront fires: Battalion Chief Jack Douglass, Battalion Chief Manuel Carter, Captain Harry Johnson, and Captain Warner Lawrence. Thanks are additionally offered to the following companies and individuals who contributed material or photographs for use in the manual:

Chicago Fire Department

W. S. Darley & Company

Industrial Chemical Products Division/3M

Mr. Scott LaGreca, Chicago Fire Department

Mr. Chris Mickal, New Orleans Fire Department

Minneapolis Fire Department
New Orleans Fire Department
Pierce Manufacturing Inc.
Wichita Fire Department

Last but not least, I thank the following staff members of Brady who played an important part in producing the final product: Mr. Matt McNearney, Fire Science Editor, who saw a need for the manual and made the necessary preparations for its production; Lynda Griffiths, who smoothed out the rough edges; and Asterisk Group, who produced the drawings that have added so much to the book.

CHAPTER 1

Fire Chemistry and Behavior

OBJECTIVE

The objective of this chapter is to acquaint the reader with the basic principles of fire chemistry and fire behavior. The various sides of the fire triangle and their relationship to fire are explored. The reader will become familiar with the cause and effect of backdrafts, flashovers, and rollovers together with the relationship of these phenomena to the fire triangle. Review questions are provided at the end of the chapter to assist the reader in determining his or her level of comprehension of the chapter contents.

Fire, electricity, and the wheel have probably had more influence on the progress of humankind than any other combination of factors. Eliminate these three items from our environment and humans would probably still be identified as nomads, moving seasonally from cave to cave in their constant fight for survival. Add to this list the elimination of the primary extinguishing agent for fire (water) and this planet would possibly be no more than a pock-marked sphere, moving through space with no signs of life aboard.

Fire has contributed its part to people's progress by providing the warmth and comfort essentially needed for our very existence. Additionally, it has produced the power required to move our vehicles and turn our machinery. Yet, when it has been on the rampage, it has turned on us with a vengeance that has been unmatched by any other enemy. This can be illustrated by the Cocoanut Grove incident in Boston in 1942 where fire snuffed out the lives of 492 people; the General Slocum disaster on the East River in New York City in 1904 where it took the lives of 1,030 people; and, even more appalling, the destruction of the Tokyo and Yokohama areas in 1923 where it snapped the lives of 91,344 people and destroyed 500,000 buildings. At other times it has chosen to destroy some of the major cities of the world, such as Rome in A.D. 64, London in 1666, and San Francisco in 1906.

In order to cope with this beneficiary that has a personality like Dr. Jekyll and Mr. Hyde, people have had to organize for the purpose of controlling and killing this potential monster whenever it chooses to vent its anger. Those who have been given the responsibility for carrying out these tasks recognize that this tyrant has the capability of seriously injuring or destroying them if they make so much as a simple mistake. Consequently, it has been necessary for these people to understand thoroughly the characteristics of fire and to be trained and capable of putting into effect the tactics required to destroy it, when necessary.

Fire chemistry and fire behavior are complex subjects; however, they are not so complex that the basic principles cannot be mastered by the average firefighter. Both are influenced by such factors as pressure, temperature, the potential fuel's surface exposure, a chain reaction, and many others. Although these variables technically have a definite impact on the genesis and progression of many fires, it is not necessary for those who are responsible for fire control and suppression to be thoroughly knowledgeable regarding all their facets. These variables can and do provide exceptions to general principles; however, the exceptions are so rare as to have minimal importance. Consequently, fire chemistry and fire behavior will be approached in this text in a simple manner that will provide firefighters and others dedicated to fire protection with the knowledge required to do an effective job on both the fireground and in fire prevention.

FIRE CHEMISTRY

In the simplest terms, *fire* might be defined as rapid oxidation accompanied by heat and flame (or heat and light). A student would normally have little trouble memorizing this and when asked the definition of *fire* he or she should be able to give a satisfactory answer; however, in reality, the student might still have little understanding of what fire really is as the definition includes the word *oxidation,* which might be foreign to him or her. *Oxidation* is a chemical process where an atom from one material combines with an atom of oxygen from another material to form a new product. An example of a product of oxidation is rust; however, in this process the oxidation takes place extremely slowly, whereas in fire the process is relatively rapid.

Most chemical processes involve the absorption or release of heat. Heat is merely a form of energy in motion. The amount of heat present or given off is referred to as *temperature.* Temperature is measured in degrees. The degrees may be stated in Celsius, Fahrenheit, Kelvin, or Rankine. Fahrenheit is generally used when referring to fires. If heat is absorbed during a reaction, it is referred to as an *endothermic* reaction. If heat is given off during a reaction, it is referred to as an *exothermic* reaction. The oxidation involved with fire is exothermic in nature. Most providers of fuel contain atoms of carbon. It is a carbon atom that attaches to oxygen atoms from another source to complete the process of oxidation in the chemical process of fire. It is necessary for a carbon atom to unite with two atoms of oxygen in order to have complete combustion. The resultant product is carbon dioxide (CO_2). Hence, it can be said that the product of complete combustion is carbon dioxide. (See Figure 1.1.)

The carbon atom will reach out and combine with a single atom of oxygen

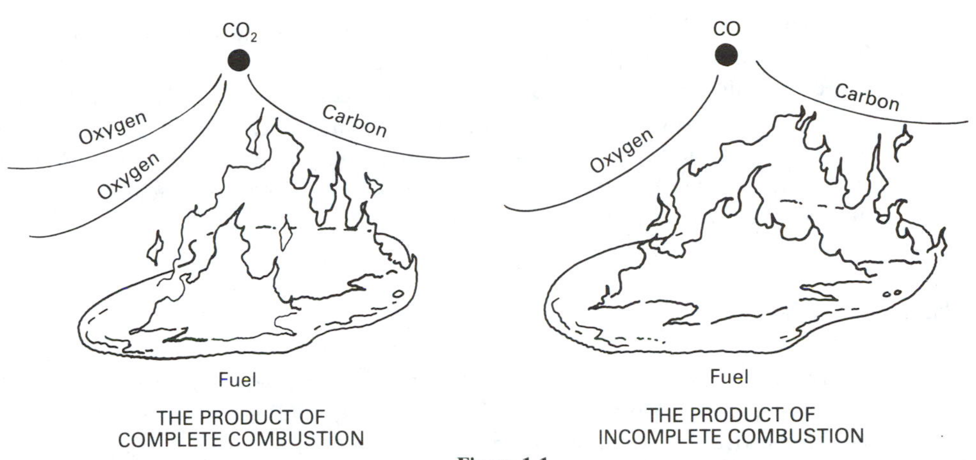

Figure 1.1

whenever there is an insufficient supply of oxygen atoms available for complete combustion to take place. In this case, the product produced is carbon monoxide (CO). (Note that the prefix *mono-* indicates one.) Hence, it can be said that the product of incomplete combustion is carbon monoxide. (See Figure 1.1.)

There are several other products of combustion, depending primarily on the composition of the material that is said to be burning. However, of the common products of combustion, the most dangerous to a firefighter is carbon monoxide. Carbon monoxide is a colorless, odorless, tasteless, and invisible gas. There are two characteristics of this gas that make it a killer.

First, carbon monoxide is a poisonous gas. Its toxicity is based on the fact that it combines readily with the hemoglobin in the red blood cells. Unfortunately, the hemoglobins that carry the oxygen through the body like carbon monoxide better than they like oxygen. As a result, they will combine with carbon monoxide 250 to 300 times more readily than they will combine with oxygen. In a very short time, a firefighter exposed to carbon monoxide may find himself or herself in a situation where the oxygen-carrying hemoglobins in his or her blood stream are so filled with carbon monoxide that he or she is no longer able to function adequately.

Ironically, this process takes place in an insidious manner, sometimes giving a firefighter little or no warning that he or she has absorbed a critically dangerous amount of the material. Although firefighters should not be working without breathing apparatus in smoky environments, if they do find themselves in such a situation, they should constantly check to see if they have absorbed too much carbon monoxide. A feeling of fatigue, a headache, or disorientation might indicate that carbon monoxide is being absorbed into the body. Absorption might also be indicated in the fingernails. The skin under the nails, which is normally pinkish in color, will turn blue as the body is deprived of oxygen. A firefighter noticing this change should immediately remove himself or herself to fresh air as he or she may be on the point of collapse and not be aware of any other signs of disablement.

The second characteristic of carbon monoxide that makes it a killer is that it is a flammable gas. In fact, it has a flammable range of approximately 4 to 74 percent. (The meaning of flammable range will be covered later in this chapter). It will ignite in a ball of flame when the proper conditions exist. This process may have been noticed by those who have seen pictures of oil tank fires, or actually had the experience of watching one burn. Frequently, large balls of fire can be seen rolling through the heavy black smoke above the tank. These balls of fire are generally carbon monoxide burning. A similar display takes place in building fires in a condition known as a backdraft. The cause and effects of backdrafts will also be discussed later in this chapter.

THE FIRE TRIANGLE

A fire occurs whenever three elements combine in the proper proportions. Basic as it may seem, these three elements (fuel, oxygen, and heat) are best illustrated by the use of the fire triangle. The three sides of the triangle are illustrated in Figure 1.2. However, to understand fully the causes of fire, it is more important to have a thorough grasp of what is meant by the proper proportions of each side of the triangle.

The Oxygen Side

Although it is possible for a fire to occur in some atmospheres other than oxygen (e.g., chlorine gas and its compounds), for practical purposes, it should be consid-

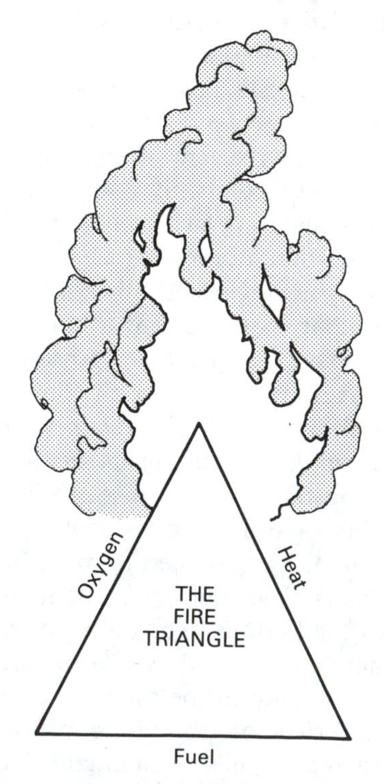

Figure 1.2

ered that oxygen is necessary in order to have a fire. Oxygen is an element that has an atomic weight of 16, which makes it 16 times heavier than hydrogen. It is essential to the burning process, but it in itself does not burn.

The primary source of oxygen for a fire is the air. The air normally contains approximately 21 percent oxygen, 78 percent nitrogen, and 1 percent other gases. If there is such a thing as a "normal" fire, it would burn best in an atmosphere of 21 percent oxygen. A fire will no longer burn normally whenever the percentage of oxygen available is above or below the 21 percent figure. The flames will begin to diminish in size whenever the concentration begins to be reduced to less than 21 percent. It will appear to die when the concentration reaches approximately 16 percent, and it will normally be completely extinguished by the time the percentage is reduced to 13 percent. Therefore, it can be concluded that a fire needs an atmosphere containing at least 16 percent oxygen to live. Ironically, the human body also requires an atmosphere containing at least 16 percent oxygen in order to survive. From this it is easy to develop an important correlation that is valuable to a firefighter. *If the atmosphere is such that a fire cannot burn, it cannot support life.* Consequently, in such situations, a firefighter should not enter the area without wearing a breathing apparatus, *and should not enter at all if additional signs indicate that the situation is ripe for a backdraft.*

Opposite to the effect of a fire diminishing whenever the concentration of oxygen is less than 21 percent is the concept that a fire will intensify whenever the supply is above 21 percent. In fact, the intensity of a fire depends on the rate at which oxygen is provided. This can be demonstrated by hitting a burning cigarette with a stream of 100 percent oxygen. The cigarette is burning normally in a 21 percent oxygen atmosphere before the pure oxygen is supplied. The pure oxygen enriches the atmosphere and the entire cigarette will be immediately consumed, from one end to the other. An example of the hazard of an enriched oxygen atmosphere to firefighters occurred at Cape Kennedy, Florida, on January 27, 1967. Astronauts Lieutenant Colonel Grissom, Lieutenant Colonel White, and Lieutenant Commander Chafee were strapped in position in the cockpit of Apollo while rehearsing for a scheduled launching when a small fire (presumably electrical) broke out in the cockpit. The atmosphere in the cockpit at the time was 100 percent oxygen. The result was the instant cremation of all three. It took 13 seconds from the time one of the astronauts said, "Fire. I smell fire" until they were dead. It is estimated that the fire may have reach a temperature of 2,500 degrees.

Air is the primary source of oxygen for a fire, but there are other sources. The most important, from a firefighting standpoint, is a group of chemicals called *oxidizers.* These chemicals contain oxygen in their makeup and release it under the process of decomposition. *Decomposition* is the breaking of a compound into its basic elements. The most common of these chemicals are nitrates and chlorates. At times these chemicals are involved in a fire situation by providing additional oxygen to a fire that is already in progress. The result is generally the development of an extremely intense, hot fire. An example might be a fire on a wooden loading dock where chlorates had previously been spilled and absorbed into the wood. The fire would intensify as the chlorates decompose and release oxygen. At other times the chemicals provide the oxygen needed to get the fire started. A good example would be a closed container of nitrocellulose film. The film itself contains an oxidizer, therefore, as it starts its process of decomposition, oxygen will be provided in a sufficient quantity to possibly complete the oxygen side of the fire triangle. If successful, a fire will be initiated without the film being exposed to any other oxygen source. Under normal circumstances, a burnable material in a closed container would lack sufficient oxygen for a fire to be initiated, even if the other elements necessary for a fire were present.

Firefighters and other people who are not aware of the capability of oxidizers to supply oxygen to a fire can get themselves in serious trouble. In 1947, a ship loaded with ammonium nitrate tied up at a wharf in Texas City, Texas. At that time, and still today, ammonium nitrate was classified as a harmless fertilizer. However, some people do not understand that it can become an explosive when it is subjected to heat, confinement, and pressure. This had been demonstrated prior to 1947 in several disasters around the world. Sometime after berthing, longshoremen discovered signs of a fire in one of the ship's holds. The crew subsequently attempted to extinguish the fire. When they were unsuccessful by using water, they battened down the hatches, covered them with tarpaulins, and blanketed off the ventilation cowls. They then turned the ship's steam fire-smothering system on in an attempt to snuff out the fire in the hold. Within a short period of time the hatch covers blew off. The fire department was notified; photographs showed that they had laid out hose lines to try to control the fire. The ship blew up approximately one hour after the fire was first discovered. All people on the wharf, including 27 volunteer firefighters, were killed. Falling embers from the explosion started a fire aboard another ship in the harbor, which also exploded approximately 16 hours later. The overall result was the death of 468 people and the injury of several thousand more. In fact, one out of every four people in the town of 15,000 were either killed or injured. A good portion of the waterfront area was destroyed with the loss estimated at $67,000,000.

The Heat Side

If you check your present surroundings, you will find that all sides of the fire triangle are present. Fuel is readily available, there is sufficient oxygen in the room, and the room is probably comfortably warm, indicating the presence of heat. However, there is not sufficient heat available to cause everything around you to burn. Ironically, however, this could happen. If the room you are in is heated to a certain point, everything in the room will break into flames, and a condition called a *flashover* will occur. The question, of course, is what is this point?

The point of reference is called the *ignition temperature*. The ignition temperature can be defined as the minimum temperature that will cause self-sustained combustion of a material, or, in plainer language, the temperature to which a material must be heated in order to burn. It should be noted that a material does not have to be in contact with the heating source in order for combustion to take place. All that is needed is for it to be heated to its ignition temperature.

Those who have been firefighters for any length of time have probably watched this process develop at a fire. Occasionally, while firefighters are laying lines to protect an exposure, the entire side of the exposure will break into flames before their eyes. The ignition temperature of the exposed material has been reached and ignition has taken place, even though the exposure was not in direct contact with the flames from the original fire. It might be added that an *exposure* is anything in close proximity to the fire that is not burning but that might start burning if some type of corrective action is not taken quickly.

On the other hand, most firefighters have at one time or another witnessed an opposite result. They have seen an easily ignitible material not ignite because the temperature of the heat source it came in contact with was too low. A good example is the spilling of a small amount of oil on an engine. The general result is a lot of smoke but no fire. This is because the temperature of the outside surface of the engine is below the ignition temperature of the oil.

Ignition temperatures vary from one material to another and have little relationship to the hazard of the material. For example, gasoline (which is considered one of the most hazardous materials associated with our daily lives) has an ignition temperature of approximately 800° F., whereas most woods have ignition temperatures in the 400 to 500° range. On the other hand, some extremely hazardous materials have low ignition temperatures. A good example is carbon disulfide, which is one of the most hazardous materials known. It has an ignition temperature in the low 200s. Agriculture dusts, which are considered extremely hazardous, have ignition temperatures in the same range as woods. Even the ignition temperature of a material appears to change over a period of time when exposed to a certain environment. In reality, however, the ignition temperature of the material does not change but the material itself changes. For instance, wood subjected to a low heat source over a long period of time will have the moisture driven out and the wood will change to charcoal. Charcoal has an extremely low ignition temperature. In fact, to prevent accidental ignition it is good practice not to subject wood to any temperature above 212° F. for any length of time.

Ignition temperatures should always be considered an approximation. Different results will be obtained from the same material depending on when it is tested, the test conditions, and the form of the material when it is tested. A different ignition temperature will be found when testing wood in small chips than when it is tested in larger pieces. The ignition temperature of gasoline will vary according to the octane rating. Regardless, the resultant principle remains the same. When all other conditions are right, a material will start to burn when its ignition temperature is reached.

Heat sources. Although the open flame is generally considered the primary heat source for a fire, there are many others.

The sun itself may be a source. It is not normally thought of as providing sufficient heat to initiate a fire, but when its rays are magnified, the resultant heat can be sufficient to raise the temperature of a material above its ignition point. A number of grass fires have occurred as a result of the sun's rays penetrating through the glass from a broken bottle. The broken bottle is shaped in such a manner to act as a magnifying glass, concentrating the resultant heat on a batch of dry grass. Fires of this type are rare, but their potential should not be ignored.

Friction is one of the oldest known sources of heat. Although early humans did not really understand the process involved, they did comprehend that sufficient heat could be developed to start a fire if they rubbed two sticks together vigorously for a long enough period of time. This knowledge proved to be a key factor in the daily survival of the tribes of nomads roving the earth. Today we have to be careful of this phenomenon in our industrial processes as we guard against slipping belts that are used to run machinery.

All the marvels of electricity carry with them the constant potential for the development of heat. The mere passage of amperage through wires causes a heat buildup. Heat can be developed to the point of raising insulation or surrounding material to above its ignition temperature whenever an attempt is made to push more amperage through a wire than the wire is capable of carrying. Additional resistance placed in the line can have the same result. The arcing that occurs from the breaking of a circuit, either by directly or accidentally opening a switch, can produce a temperature of over 2,000 degrees. Similar temperatures can be found in the sparks caused by a crossed circuit. Of course, nature's sources of electrical current also result in temperatures that can ignite combustible mixtures. The two most common are static electricity and lightning. Safeguards are built into hazardous areas such as operating rooms to protect the environment against the potential danger of static electricity, and additional devices are used to protect entire buildings from lightning when the heavens go on a rampage.

When compressed, gases develop heat. This is referred to as the *heat of compression.* Extreme temperatures can be developed in this manner. A good example where this principle is used in a beneficial setting is the diesel engine. Air alone is compressed on the compression stroke. The temperature of the compressed air reaches well above the ignition temperature of the fuel oil used in the engine. The oxygen side and the heat side of the fire triangle are then present inside the cylinder. The fuel oil is injected into the cylinder when ignition is desired. Burning commences immediately.

Dangerous levels of heat can also be produced by mixing two chemicals. Some reactions are slow; others are instantaneous. A few years ago a number of small fires were reported in drugstores in a large metropolitan area. These fires were started by school children who had learned in a chemical class that heat could be developed by mixing a certain type of hair cream with another common product found in drugstores. They would enter a store, mix the two materials together on a shelf, and leave. A small fire would break out a short time later.

A much faster reaction takes place when other types of chemicals are mixed. A good example is the propellant system used in some rockets. Two different chemicals are carried in separate storage tanks. Controlled mixing of the two occurs in the combustion chamber. The mixing of the two causes instant ignition, completely eliminating the need for another ignition source in the rocket. This type of response is referred to as a *pyrophoric reaction.*

Another source of heat is caused by a process referred to as *spontaneous ignition.* Generally, spontaneous ignition is caused by oxidation. Those combustible materials that are capable of combining with oxygen will, in most cases, become involved in the oxidation process when they are exposed to the air. The result of the oxidation is heat generation; however, the heat is usually dissipated as fast as it is developed. Materials that are capable of spontaneous ignition, however, under the proper conditions, build up heat faster than it is lost to the surroundings. The time required for the heat to build up to the point where ignition takes place will vary with the condition under which the heating takes place. It will also vary from one material to another and with the same material at different times, depending on its moisture content and other factors. As an example, spontaneous ignition may take place in hay within a period of two to six weeks after it is placed in storage.

Heat movement. Most fires in combustible materials are started when the burnable material comes in direct contact with a heating source. In such cases the heating source is above the ignition temperature of the burnable material. The most prevalent type of direct contact is the open flame, with the match playing a big part. A match burns at a temperature of approximately 2,000° F., which is well above the ignition temperature of most combustible materials. When held in contact with a burnable material, it will raise the material to its ignition temperature unless the heat is dissipated within the material faster than it is being formed at the contact point. This same principle holds true regardless of the heating source. The heat buildup at a contact point must be faster than it is being dissipated to the surroundings if ignition is to occur.

There are three heat sources other than direct flame contact that can raise the heat level of a material to above its ignition temperature. The three are radiation, conduction, and convection. These are methods by which heat moves from one location to another. They are samples of nature's attempt to maintain a heat balance by movement of heat from a warmer to a colder body. (See Figure 1.3.)

Radiation. Heat is radiated from a burning fire in all directions. Some of the radiated heat is transferred back into the fire and helps accelerate the chain-reaction process. The rays travel in a straight line and would travel at the speed of light if they were moving in a vacuum. Radiated heat is a problem at all fires. The rays travel in any direction and continue to move in that direction until their energy is dissipated or they are stopped by an intervening body. It is generally radiated heat that penetrates the exposures in single-story structures and ground-level open fires. When streams from lines laid at fires are used to wet down adjacent structures, they are so used to prevent the radiated heat from the fire from raising the structures to above their ignition temperatures.

Conduction. The second method of heat moving from a source to an exposure is called conduction. The principle of conduction is simple. When two materials having different temperatures are placed in contact with one another, the material having the greatest heat will transmit some of it to the material having the lowest heat. The temperatures of the two materials will become identical over a period of time.

The principle involved can be demonstrated by a person holding a metal rod with the opposite end of the rod placed in an open fire. The rod will become hot enough after a period of time that the person will have to drop it. What happens is that all the molecules within the rod are in contact with other molecules. Those in

Figure 1.3 The three methods of heat movement

the fire start transmitting heat to adjacent molecules and movement continues up the rod until the person holding it senses the heat. The principle of conduction can cause fires to start inside walls of structures, but it plays a more important role when the structures are made of metal. A good example is a fire aboard a ship. A fire in one hold can be transmitted to the cargo in another hold by heat conducted from one hold to the other through the metal bulkheads. A fire in a stateroom can be transmitted to an adjacent stateroom. In fact, it can move in one of six directions, or perhaps in all six at the same time—to the areas adjacent to the four bulkheads, to the area above, and to the area below.

Convection. The third method for transferring heat from one location to another is called convection. Convected heat moves in currents. The currents move in an upward direction from the fire. They continue in an upper direction until stopped by a ceiling or other obstruction. Then they move along the ceiling, searching for another vertical opening. If they find one, they will again start in an upward direction. If they do not find a vertical opening, they will move until they hit a wall, and then start a downward movement. The vertical rise of the heat and products of combustion, their banking along the ceiling, and their travel down the walls is referred to as the *mushroom effect.*

Convected heat is the culprit in multiple-story fires. The heat will generally continue in an upward direction as long as it can find a vertical opening. A vivid example of its movement was demonstrated in a five-story hotel fire in the Los Angeles area. The fire started in a room adjacent to an unprotected stairwell on the third floor. The heat from the fire left the room and traveled up the stairwell to the top (fifth) floor. It spread across the entire length of the hall on the top floor and through open transoms into the rooms. Seven people died on this floor. It spread approximately halfway down the length of the hall on the fourth floor. Damage from the natural spread on the third floor was limited to approximately a fourth of the length of the hall area.

Although convected heat is normally thought of as moving in an upward direction, its path of travel can be changed by the wind or drafts. Many times the drafts are induced by firefighters in their ventilation procedures. A poor choice for cutting a hole in a roof can cause the fire to be transmitted to uninvolved areas of the building through horizontal openings.

The Fuel Side

Most people who first cast their eyes on the fire triangle form the opinion that the fuel side would be the most easily understood; however, they are generally mistaken. They have little problem conceptualizing that fuel comes in three forms—solids, liquids, and gases—but this concept is erroneous and can cause a certain amount of confusion. With the exception of certain metals, solids and liquids do not burn. Both of these give off vapors when subjected to heat. It is the vapors that burn and not the material itself. Therefore, they should not be considered as fuel but rather as fuel providers. This principle can be observed by watching a candle. Note that the flame does not actually touch the wax. What happens is that the heat from the flame reduces the wax to a vapor. The vapor rises and burns. This same phenomenon can be observed by taking a close look at a fire in a fireplace. From a distance the flames appear to be in contact with the wood, but a closer look will show that they are not. What is happening is that the vapors given off by the wood are burning a short distance above the wood. For practical purposes, then, it can be said that only vapors are involved in the burning process.

But all vapors do not burn. This becomes apparent on those occasions when a person tries to start his or her car and it will not cooperate, even though there are vapors in the cylinder and an adequate spark for ignition is being provided. At times there are insufficient vapors in the cylinder to burn and at other times the concentration is too great. Whenever there are insufficient vapors available, the mixture is said to be too "lean" to burn. When there is an excessive amount of vapors, the mixture is too "rich" to burn.

These examples illustrate the principle that vapors will burn only when conditions are right. They will burn within certain limits, which are referred to as the *flammable* (or *explosive*) *limits.* The flammable limits are expressed as a percentage of a vapor/air mixture. For example, if the flammable limits of a certain flammable liquid are 2.0 to 8.0, it means that the vapors will burn only when the vapor-air mixture is between 2.0 percent vapor and 98 percent air, and 8.0 percent vapor and 92 percent air. To understand this concept thoroughly, it is necessary to be familiar with the evolution process that takes place as a flammable liquid changes from a liquid to a vapor.

Figure 1.4 illustrates an open cup flash point tester. Open cup testers are used for some liquids, and the closed cup tester is used for others. Different flash points will be obtained for the same liquid if tested by both the closed cup and open cup method; however, the concept of the testing and the principles involved is the same for both instruments.

A thermometer has been placed in the liquid to be tested. It is used to determine the temperature of the liquid at two points in the testing process. The spark plug above the liquid is constantly providing a spark that has a temperature above the ignition temperature of the liquid. The flame below the cup is slowly raising the temperature of the liquid. At some point, a flash will occur across the face of the liquid and go out. The temperature at which this occurs is called the *flash point.* Heating of the liquid is continued and a few degrees above the flash point (usually 2 to 4) a flash will occur above the liquid and the resultant fire will continue to burn. The temperature at which this occurs is called the *fire point.*

Examine what happened. Vapors were being given off from the liquid before the first flash occurred; however, the mixture formed by these vapors with the air would not burn. The mixture was too lean. (See Figure 1.5.) When the first flash occurred it indicated that the vapor/air concentration was burnable. Therefore, the flash point was the temperature at which the vapors first entered the flammable range. The flash across the top of the liquid burned off all the available vapors but more were developed as the temperature of the liquid continued to rise. At the second flash (fire point) the vapors were within the flammable range and were being given off at a sufficient rate to continue the burning.

Figure 1.5 will put all this in perspective. Remember that it was stated earlier that all sides of the fire triangle had to be in the proper proportions in order to have a fire. The proper proportion for the fuel side of the triangle is for the fuel/air vapor from the material that will be said to burn to be within the flammable range.

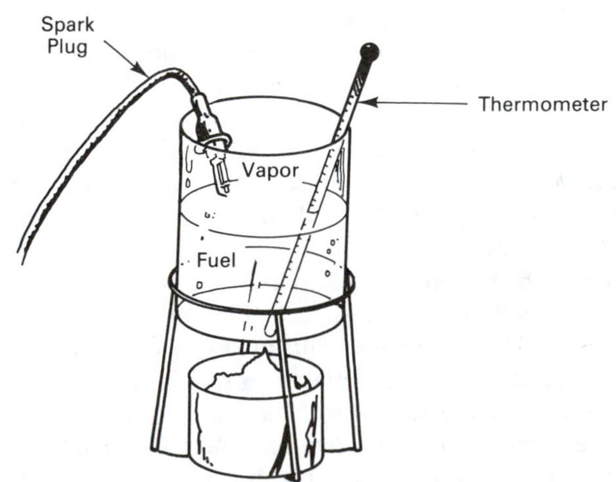

Figure 1.4 An open cup flash point tester

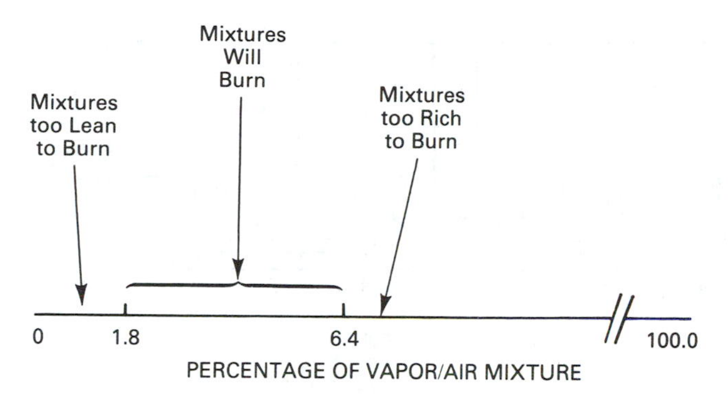

Figure 1.5

To illustrate, the flammable liquid shown in Figure 1.5 has a flammable range from 1.8 to 6.4. Burning will not take place (the mixture will be considered to be too lean to burn) when the concentration of vapors is below 1.8. Nor will burning take place when the concentration is above 6.4 (in this case the mixture will be too rich to burn). Burning can occur only when the vapor/air concentration is between 1.8 and 6.4.

The flammable limits for both liquids and gases are measurable and vary from one product to another. They also vary with the ambient temperature. The limits for some of the common liquids and gases at 70° F. are:

Acetylene	2.5–81.0
Butane	1.9– 8.5
Carbon monoxide	12.5–74.0
Gasoline	1.4– 7.6
Kerosene	0.7– 5.0

The flammable range of liquids and gases can be measured, but the burnable limits for gases given off from solids cannot. It is assumed, however, that these gases also only burn within predetermined limits and that the principle involved is the same as that of measurable fuels.

FIRE BEHAVIOR

Fires generally start small and continue to grow until one of the sides of the fire triangle begins to disappear. The fire will go through two or three definite stages during this process. These stages are defined as the incipient stage, the free-burning stage, and the smoldering stage. (See Figure 1.6.) Following are some brief descriptions of the three stages. When considering the three stages, it should be kept in mind that they are defined only for the purpose of understanding the general progress of a fire. Under actual conditions, there is no clear line of demarcation from one stage to another.

INCIPIENT STAGE—
Fire just starting

FREE-BURNING STAGE—
Fire spreading rapidly; heat accumulating at upper levels; diminishing oxygen content in room

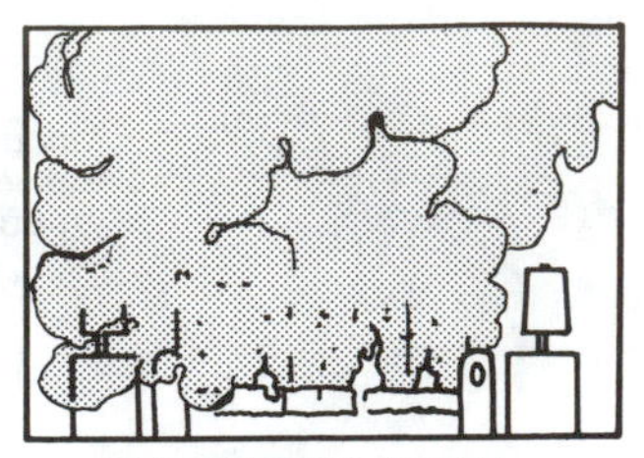

SMOLDERING STAGE—
Flickering flames; heavy smoke; oxygen content below 16 percent

Figure 1.6 The three stages of a fire

The Incipient Stage

This is the beginning stage. The fire will quickly start consuming more material as the chain-reaction process takes place. The gaseous products of combustion, such as carbon dioxide and carbon monoxide, will move in an upward direction with the smoke produced. The temperature of the fire will rise to above 1,000° F. and the room will begin to heat up. The oxygen content during this phase will remain at or near 21 percent. A short time later the fire will move into the second stage.

The Free-Burning Stage

Temperatures begin to increase rapidly during this stage if sufficient oxygen is available. The flames increase in size and intensity. If the fire is in an open area with sufficient fuel and sufficient oxygen, it will continue in this stage and maintain its growth until positive action is taken to stop it. The oxygen concentration remains at or near 21 percent and the products of combustion are carried upward and away from the fire. However, if the fire is in a closed area the temperature at the ceiling will increase to above 1,300° F., the products of combustion will begin to build at ceiling level and start banking down the walls, and the oxygen content in the room will start diminishing as conditions approach the third stage.

The Smoldering Stage

As the fire continues into the free-burning stage in a closed room, it will use up more and more of the available oxygen. Carbon monoxide will form a greater part of the gases of combustion and flame propagation will start diminishing in size. The flames will have been reduced to a flickering stage when most of the oxygen in the room has been consumed. At this point the oxygen content is at or below 16 percent. The flames may disappear completely and the fire will appear as nothing but a glow if allowed to continue in this condition for any length of time. In this situation, there is generally still sufficient fuel in the room to burn and the entire room will be in a super-heated condition.

Photo 1.1 This fire has adequate oxygen to continue in the free-burning stage. (Courtesy of Chris Mickal, New Orleans Fire Department Photo Unit)

THE PHENOMENA OF THE FIRE TRIANGLE

There are three phenomena of the fire triangle that should be thoroughly understood by every firefighter. The cause of each is different, but the overall result is almost identical—*all are potential killers.*

Backdrafts

The first phenomenon is referred to as a backdraft. A backdraft occurs when the fuel side and the heat side of the triangle are in proper proportions in an area and the oxygen side is suddenly supplied. A typical example is a tight room in which the fire has been burning for some time. It has reached the smouldering stage and there is insufficient oxygen in the room for burning to continue. The gases in the room are super-heated. In fact, the temperature in the room is above the ignition temperature of the materials in the room. If a door is opened into the room, air will rush in and complete the fire triangle. The air mixes with the carbon monoxide in the heavy smoke, resulting in a large ball of flame. The overall effect resembles an explosion, with the explosive force normally leaving the room through the same entry that was used by the air. The person who opened the door and entered the room would probably be killed or seriously injured. (See Figure 1.7.)

The time it takes for a backdraft to occur once oxygen is provided into an

Figure 1.7 A backdraft condition—The room contains fuel and heat. Opening a door provides the oxygen.

area varies according to the size of the area where the fire is located. The reaction is almost instantaneous when the fire area is small. In large areas, it may take as much as two minutes for a sufficient amount of oxygen to enter the area to complete the fire triangle. A couple of examples will illustrate this point.

A fire occurred in a small closet at the end of a hall in a dwelling. It had burned for a sufficient period of time to reach the smoldering stage. The hall stretched from the front door of the dwelling to the closet. The firefighters took a booster line in through the front door and advanced to the closet. The nozzleman opened the closet door without first checking for signs of a potential backdraft. The backdraft was instantaneous. The resultant explosion blew all three firefighters down the hall, out the front door, across the lawn and rolled them onto the front yard of the house across the street. Fortunately, no one was seriously injured.

Another fire occurred in a large manufacturing warehouse. Upon arrival at the scene, the company officer went inside to size-up the situation. He found a small fire at the back of the warehouse. He returned to the apparatus and reported to the dispatch center that he had a small fire that he could handle. He instructed his crew to take a booster line inside and put out the fire. When the crew entered the building, they found the entire warehouse involved with fire. A backdraft had occurred in the officer's absence. He went back to the apparatus and ordered a second alarm assignment.

At another backdraft situation the firefighters were not as fortunate. Two salvage company members had entered a large warehouse for the purpose of covering the stock. They had been throwing covers for approximately two minutes when they saw the backdraft coming. The first signs they had were flashes of blue light resembling a neon light at ceiling level. One of the firefighters dropped and covered his head with his turnout coat. The other tried to run. The explosion went over the head of the man who had dropped but the firefighter who ran was killed.

Although signs of a potential backdraft are not always present in large areas, in small areas they generally are. Before entering any closed area where a fire is burning, a firefighter should check the door to see if it is hot. If it is, he or she should not enter the room. This is especially true if air is being sucked into the room or small puffs of smoke can be observed coming out from under the door. The small puffs of smoke are caused by air entering the room under the door and backdrafting out.

If possible, top ventilation should be started before the room is entered whenever a potential backdraft condition is encountered. Top ventilation is the creation of an opening above the fire to allow the heat and smoke to escape. (Information on top ventilation will be provided later on in the text.) If top ventilation is not possible, it may be practical to cut a small opening into the room and stick a spray nozzle through the hole. Swirl the nozzle or direct the stream to the ceiling level until the temperature of the interior of the room has been reduced below the ignition

Photo 1.2 Firefighters should always wear breathing apparatus and be on the alert for a backdraft when entering a smoke-filled area. (Courtesy of Chicago Fire Department)

temperature of the material inside. One indication that this has taken place is a reduction in the generation of steam.

In summary, backdrafts are potential killers. Always be on the alert for signs indicating their potential. Some firefighters get careless because so many of their routine fires resemble a backdraft situation. Heavy black smoke is usually present inside a building and heat is always there. The heavy smoke and heat are also present in a potential backdraft; however, the elements are in the proper proportion to complete the triangle. Always check. It is much smarter to assume that conditions are ripe for a backdraft than assume the opposite.

Flashovers

The second phenomenon is referred to as a flashover. The two sides of the fire triangle that are present in the proper proportions immediately before a flashover occurs are the oxygen side and the fuel side. Note that this is not any different than what occurs when a match is used to start a fire in a small pile of papers. The dissimilarity is that with the papers the fire starts small and relatively slowly increases in size, whereas with a flashover a large area becomes involved in fire instantaneously.

Perhaps the best way to visualize the process is to imagine a room approximately 12 feet by 12 feet in size in which there is sufficient oxygen for a fire and sufficient fuel. If the entire room is slowly heated, the temperature in the room will eventually reach the ignition temperature of everything in the room, and the total contents will start to burn at the same time. It would then be said that the room "flashed over." The overall result would be similar to a backdraft. Any one who happened to be in the room at the time would probably be killed or seriously injured. (See Figure 1.8.)

In addition to different sides of the fire triangle being supplied to cause a backdraft or a flashover, there is another difference that contributes to the hazard. Generally, warning signs are present whenever an area is ripe for a backdraft. This is not true with a flashover. Flashovers do not provide any positive telltale signs and therefore occur with little or no warning. They take place in areas in which firefighters find themselves all the time. These are areas where there is oxygen, fuel, and lots of heat. However, in our day-to-day firefighting, the heat at the floor level in an area is normally well below the ignition temperature of the contents of the room.

In flashover situations, heat and smoke from a fire in a building will travel to other rooms and areas of the building. The heat will start building up at ceiling level, sometimes in a small room and sometimes in large areas. The heat buildup is generally above the heads of firefighters who are down low, advancing lines. A flashover will occur when the temperature of the material at ceiling level reaches the ignition temperature. The first instantaneous flash will generally be above the heads of the firefighters; however, it will quickly spread to lower levels due to the tremendous volume of heat involved.

It is difficult to establish any positive tactics to protect firefighters from flash-

Figure 1.8 A flashover potential—The room contains fuel, heat, and oxygen. The heat at the ceiling level is close to the ignition temperature of combustibles in the room.

overs, since one of the early principles they are taught is not to open up a hose line until they see the fire. The reason for this is to try to keep water damage to a minimum. However, when firefighters find themselves in a situation that is clearly on the edge of a flashover, the flashover might be stopped if spray streams are used momentarily at ceiling level to reduce the heat buildup.

Rollovers

A third phenomenon of the fire triangle is referred to as a rollover. The phenomenon is called this because the resultant fire manifests itself in a rolling motion, normally at ceiling level. It is caused by the smoke at the ceiling level suddenly igniting as a result of being heated to its ignition temperature and mixing with sufficient oxygen to complete the fire triangle. A typical case follows. (See Figure 1.9.)

Firefighters have advanced a line to a room on fire. The door to the room may have been left partially open, allowing smoke to enter the hallway or allowing smoke to escape from the room and start moving down the hallway as the firefighters crack open the door. They are down low with their line charged but the nozzle closed. When they crack open the door they see that the room is filled with heavy black smoke. They push open the door—prepared to enter. Suddenly the

Figure 1.9 A rollover condition—The fire rolls out of the door, over the heads of the firefighters, and down the hall.

smoke over their heads and in the hall breaks into flames and appears to roll down the hallway. They open their line and attack the fire.

Rollovers occur more often than do backdrafts or flashovers. They can be killers if firefighters enter the room standing up or if their line is not charged when the rollover occurs. To protect themselves against this phenomenon, firefighters should make it a practice to enter fire areas crouched down low with lines charged. If they feel that the rollover is a threat to their safety, there should be no hesitation in opening their line and directing it at the ceiling level in a spray form.

REVIEW QUESTIONS

1. In simple terms, what is the definition of fire?
2. What is oxidation?
3. Give an example of slow oxidation.
4. What is heat?
5. What is temperature?
6. In what units is temperature measured?
7. Give four different units in which degrees are measured.
8. What unit measurement of degrees is generally used when referring to fires?
9. What is an endothermic reaction?
10. What is an exothermic reaction?
11. Is the oxidation involved with fire an endothermic or an exothermic reaction?
12. What atom attaches to oxygen atoms in the chemical process of fire?
13. How many oxygen atoms unite with a carbon atom when there is complete combustion?

14. What is the product of complete combustion?
15. What is the product of incomplete combustion?
16. What is the most common product of combustion that is dangerous to firefighters?
17. What are the two characteristics of carbon monoxide that make it a killer?
18. Upon what factor is the toxicity of carbon monoxide based?
19. How much more readily will the hemoglobins in the blood stream combine with carbon monoxide rather than oxygen?
20. What happens when the hemoglobins in the blood stream begin to fill with carbon monoxide?
21. What method can a firefighter use to check to see if he or she has taken in too much carbon monoxide?
22. What is the flammable range of carbon monoxide?
23. What are the three sides of the fire triangle?
24. What is the atomic weight of oxygen?
25. How flammable is oxygen?
26. What is the primary source of oxygen for a fire?
27. How much oxygen is there in the air?
28. What happens to a fire when the oxygen content falls below 21 percent?
29. What percent of oxygen is needed in the air in order for a fire to live?
30. How much oxygen is needed in the air in order for a person to live?
31. Upon what does the intensity of a fire depend?
32. From a firefighting viewpoint, what is the most important source of oxygen other than the air?
33. What is decomposition?
34. What are the two most common chemical oxidizers?
35. What would probably be the result if oxidizers suddenly become involved in a fire in progress?
36. What are the three elements that can cause ammonium nitrate to become an explosive?
37. Define ignition temperature.
38. What causes an exposure to break into flames?
39. What is an exposure?
40. What is the general relationship between the ignition temperature and the hazard of a material?
41. Approximately what is the ignition temperature of gasoline?
42. What is the approximate ignition temperature of most woods?
43. How accurate are ignition temperatures given in charts?
44. What is generally considered the primary heat source for a fire?
45. How can the sun be considered a heat source for a fire?
46. What temperature can be produced by the arcing that occurs as a result of breaking an electrical circuit?
47. What are the two most common sources of electrical current?
48. What is meant by the "heat of compression"?
49. Give an example where the "heat of compression" is used in a beneficial setting.

50. What is a pyrophoric reaction?
51. What is the general cause of spontaneous ignition?
52. At approximately what temperature does a match burn?
53. By what three methods (other than direct flame contact) can a material be raised to a heat level above its ignition temperature?
54. What is meant by radiated heat?
55. What is the speed of heat radiation?
56. What is meant by conducted heat?
57. What is the principle of conduction?
58. What is convection heat?
59. In what type of structures does convection play a big part?
60. In general, do solids and liquids burn?
61. What term is used to identify a situation where there are insufficient vapors to burn?
62. What term is used to identify a situation where there is an excessive amount of vapors and burning will not take place?
63. Vapors will burn only within certain limits. What are these limits called?
64. What does the 4.1 refer to when it is said that the lower flammable limit of a combustible liquid is 4.1?
65. In the testing of a flammable liquid in an open cup tester, a flash will occur across the top of the liquid and go out. What is this point called?
66. A few degrees above the point referred to in question #65, a flash will occur and the fire will continue to burn. What is this point called?
67. What are the flammable limits of gasoline?
68. Do the gases given off from solids have a flammable range?
69. What are the three stages of a fire?
70. What is meant by the incipient stage of a fire?
71. What is meant by the free-burning stage of a fire?
72. Explain the smoldering stage of a fire.
73. What side of the fire triangle is provided in order to have a backdraft?
74. How long after oxygen enters a room does it take for a backdraft to occur?
75. What are some of the signs of a potential backdraft?
76. If possible, what type of ventilation should be started before entering a room where there is a potential backdraft condition?
77. What should be tried when a potential backdraft is encountered and top ventilation is not possible?
78. What side of the fire triangle is supplied when a flashover occurs?
79. What is one of the big differences between a potential backdraft situation and a potential flashover situation?
80. How might a flashover be stopped if it is recognized that one is imminent?
81. What is a rollover?
82. What is the best way for firefighters to protect themselves against the dangers of a rollover?

CHAPTER 2

Fireground Planning and Coordination

OBJECTIVE

The objective of this chapter is to acquaint the reader with the various facets that are considered in fireground planning and coordination. The functions for which the fire department is responsible on the fireground are identified. An explanation is provided as to whom these functions are assigned in order to achieve maximum efficiency on the fireground. Review questions are provided at the end of the chapter to assist the reader in determining his or her level of comprehension of the chapter contents.

Between June 3 and June 7, 1942, land- and carrier-based planes attacked the Japanese fleet northwest of the island of Midway. In the opinion of many military experts, the U.S. victory in this battle was the turning point in the Pacific campaign of World War II. The Battle of Midway was the first Japanese defeat. Not only did it end Japan's thoughts of seizing Midway Island and using it as a base to attack the Hawaiian Islands but it crippled Japan's naval air power. However, the battle was not completely onesided. American naval forces also had their share of casualties. One of the situations that occurred during the battle will probably go down in history as the most disastrous defeat of a naval aircraft squadron. Every plane from Torpedo Squadron Eight was shot out of the air, and every pilot and crewman—except one—was killed. The destruction of Torpedo Squadron Eight while making an attack on a Japanese carrier will remain a prime example to tactical commanders of all emergency forces of the tremendous losses that can occur when a planned attack lacks coordination.

During World War II, a naval torpedo squadron was one of three squadrons required to make a successful coordinated attack on an enemy carrier. The other two squadrons were the fighter squadron and the dive bomber squadron. Coordination required that planes from all three squadrons commence their attack at the same time. The fighter planes would come in at an angle and strafe the deck, hoping to clear it of firepower. The dive bombers would approach from overhead and release their bombs while the torpedo planes would launch their attack at water level to drop torpedoes. This attack was generally successful, if properly coordinated. Unfortunately, on the day of the conflict, planes from Torpedo Squadron Eight arrived on the battle scene prior to the arrival of the fighter and dive bomber squadrons. The planes were low on fuel, having just barely enough to make an attack and return to their own carrier. The commander of Torpedo Squadron Eight decided that he could not wait for the arrival of the other two squadrons, and ordered the attack. The outcome is history.

The necessity for coordination on the fireground is just as important as it is for a military operation. Although the elements required to be completed and coordinated differ in the two situations, the result of failure is the same. Either the battle will be lost or the losses during the battle will be greater than desired.

To understand the importance of coordination on the fireground, and the manner in which coordination is achieved, it is first necessary to be familiar with what operations are required in order to extinguish the fire with the least possible loss of life and property. It is also necessary to understand what is required to carry out an operation successfully, and who is responsible for seeing that the operation is properly completed.

It has often been said that no two fires are exactly alike. However, it has also been said that all fires are somewhat alike. Because there is a degree of similarity in all fires, thought must be given to certain factors regardless of what type or size of occupancy in which the fire occurs. Consideration must also be given to the tasks to be performed on the fireground, and the manner in which these tasks can best be coordinated.

Coordination on the fireground requires proper timing; it also requires knowledge on the part of every individual involved as to the exact part he or she plays. Assignments of the tasks to be performed must be made in advance, and not left to the snap judgment of a field commander at the scene of an emergency. Consequently, successful operations at a fire demand that the tasks to be performed be divided and assigned long before the bell rings. To do this, it is first necessary to review the tasks required and ensure that those individuals responsible for their performance are properly trained. The importance of this might be related to an article written by a famous football coach many years ago and published in the *Saturday Evening Post*. The essence of the message in the article was summarized in the title: "Football Games Are Not Won on Saturday."

Following is a list of the common tasks that must be performed on the fireground. They are not listed in priority order but rather in a format that will simplify their division later on in the chapter. With the exception of the size-up, which is performed for the most part by the officer-in-charge of the fire, each of the functions will later be allocated to a company to ensure that the responsibility for its execution is identified. Not all of the tasks will have to be performed at every fire, but common sense dictates that thought should be given to each individual task prior to its abandonment. The common factors that must be considered are:

Size-up
Fire extinguishment
Protection of exposures
Laddering
Overhaul
Emergency medical care
Forcible entry
Physical rescue
Control of utilities
Salvage
Ventilation

THE SIZE-UP

It would be very difficult, if not impossible, to intelligently and successfully conduct operations at the scene of an emergency without first making an estimate of existing conditions. By necessity, existing conditions include available personnel, equipment, water supply, life hazard, time of day, weather, type of occupancy, what is burning, and size of the fire, just to name a few. This estimate of existing conditions is referred to as the *size-up*.

Although the tactical size-up for overall operations at the fire is the responsibility of the officer-in-charge of the fire, every person at the fire, from the newest rookie to the most senior fire officer, consciously or unconsciously participates to some degree in some portion of the size-up process. Consequently, the more knowledgeable a firefighter is regarding the size-up process, the more likely he or she is to make a good decision regarding his or her actions on the fireground.

The size-up commences long before the alarm sounds and continues throughout the duration of the emergency. For practical purposes, it can be divided into three parts—the pre-alarm size-up, the response size-up, and the fireground size-up.

The Pre-Alarm Size-Up

The pre-alarm size-up first manifests itself when any information on the fire building or its exposures that could affect operations on the fireground is initially gathered. This might have taken place on a previous response to the occupancy; however, it generally occurs during an inspection of the fire building, which is made prior to the fire occurring. This inspection is a portion of an overall process referred to as *pre-fire planning*.

Pre-fire planning. The objective of pre-fire planning is to fight the fire before it occurs. It involves collecting information and using the information to plan ahead of time how fires will be fought if they occur in various parts of the building. The pre-fire planning inspection can provide a majority of the information required to make a successful analysis of the situation on the fireground. During this inspection, information should be obtained on the following:

1. *The size and construction of the building.* Knowing this information will help determine whether handlines can be used or whether it may be necessary to resort to heavy stream appliances. It should be remembered that the reach of a hand-held hose line is about 50 feet. It is possible for the fire to burn unchecked in the center of the building if it is attacked from both sides with handlines if the width of a building exceeds 100 feet. (See Figure 2.1.)

The hazard of this was brought to the attention of the entire fire service during the early 1940s with the destruction of the General Motors plant in Livonia, Michigan. This fire occurred during the daytime when the plant was in full operation. Attacks were made from both sides of the structure; however, the plant was lost because the streams did not have the reach to knock down the fire in all parts of the building.

The construction of the building also plays a major part in making a decision as to how the fire should be attacked. Although a building of concrete construction may prove to be beneficial in limiting external fire spread, it may also make access to the fire much more difficult due to the large amounts of gases, smoke, and heat that will be retained within the building.

In some cases, such as in high-rise buildings, the interior of the building may become like a heating oven due to the entrapment of the fire, heat, and smoke. It is estimated that the temperature inside in the First Interstate Bank Building in Los Angeles in May of 1988 reached approximately 2,000 degrees, due to the construction of the building.

Buildings constructed of unprotected steel may prove to be even more of a problem. The steel will begin to lose its strength in as short a period of time as 10 minutes when subjected to severe fire conditions, The size-up should consider the

Figure 2.1 Consider the reach of streams

Photo 2.1 The size and construction of a building play an important part in fire extinguishment. (Courtesy of Chris Mickal, New Orleans Fire Department Photo Unit)

possibility of early collapse of the building if the fire has burned unchecked for any period of time.

Additionally, the construction of the building could have an adverse effect on various related factors such as access, ventilation, salvage, and so on.

2. *The life hazard.* Saving a life takes first priority at every fire. (See Figure 2.2.) An aggressive attack on the fire may be the best method of saving life under practical conditions, but thought should always be given to the possibility of people being trapped. Consequently, every possible condition that may contribute to loss of life during a fire should be considered during the pre-fire planning inspection. How many people are likely to be in the building at the time of the alarm? Will all of them be able to leave the building unassisted? Is it likely that people might become trapped above the fire floor, which may require the use of helicopters to rescue them? Is there an adequate number of exits in the building? Are the exits likely to be blocked by fire during the emergency? What will be the best way to get to people in various parts of the building, and what will be the best way to get them out?

3. *Fire and smoke travel.* Consideration must be given during the inspection as to how the fire, heat, and smoke will most likely travel to various parts of the building. Evaluation of conditions should include thought given to heat traveling

Figure 2.2 Rescue has first priority at a fire

by radiation, convection, and conduction, and the possibility of fire traveling from room to room, from floor to floor, and from the fire building to other structures or material outside of the building. When considering this factor, it should be remembered that fire can travel from room to room on the same floor:

- **a.** Through unprotected horizontal openings such as doorways, hallways, interior windows, transoms, breaks in walls, and so on.
- **b.** By convection of heated air, smoke, and gases.
- **c.** As a result of backdrafts or flashovers.

d. By the conduction of heat through metal pipes, other metal objects, or possibly through walls.
e. Through concealed spaces, air-conditioning systems, and the like.

Fire can travel from floor to floor:
a. Through unprotected vertical openings such as open stairwells, elevator shafts, light wells, and so on. The fire itself may travel through these openings or the extension of the fire may be caused by convection heat.
b. By burning through the floor or by conduction through the floor.
c. Through windows, by fire extending up the outside of the building and through the windows of upper floors. This is particularly a problem in high-rise buildings.
d. By sparks or burning material falling to lower floors.
e. By roof or floors collapsing onto lower floors.
f. Through air-conditioning and similar systems.

Fire may extend from building to building, or from the fire building to material outside of the building:
a. By conduction, convection, or radiation.
b. Through unprotected wall openings.
c. Through combustible walls or roofs.
d. By failure of walls through explosions or structural weaknesses.
e. By flying brands landing on combustible roofs of buildings.

4. *The contents of the building.* Knowledge of the contents of the building is extremely important when assessing the speed with which the fire can be expected to travel. If flammable liquids are stored in the building, it can be expected that the fire will spread rapidly if they become involved. Rapid expansion of the fire can also be expected if material such as cotton is involved. More than having an effect on the fire spread, what is stored inside the building may have an adverse affect on floor collapse. Thought should be given not only to the weight of the material in its stored condition but also to its potential weight when it absorbs water.

5. *On-site fire protection.* The location of all on-site fire protection systems should be identified. All shutoffs and fire department inlets should be located if the building has a sprinkler system. The areas controlled by shutoffs and inlets should be recorded. Particular attention should be paid to any parts of the building that are not covered by the sprinkler system. If the occupancy is of sufficient size to warrant an annunciator or central control station that will identify the location within the building where a sprinkler head is flowing water or an alarm box has been pulled, then this location should be noted for future reference. The time required to find the seat of the fire will be shortened if responding companies are familiar with the location of these central controls. Controls for carbon dioxide, halon, and other fire protection systems should also be identified and their areas of protection noted.

If the building is five or more stories in height, it probably will be equipped with a dry standpipe system for fire department use. This system is helpful when relaying water to upper floors. The inlets to this system will most likely be found alongside those for the sprinkler systems; if they are not, however, they should be located during the pre-fire inspection. Of course, it is important that an assessment be made of the length of lines that will be required to be laid into all the inlets of all systems.

6. *Ventilation problems.* The pre-fire planning inspection is the ideal time to make an analysis of the best methods for ventilating the building during the fire and the ventilation problems that are likely to arise. It is generally best to start the inspection of the building on the roof. Not only does this location provide an opportunity to size-up the external exposure problem but it also gives firefighters an opportunity to determine the best means of providing roof ventilation. While on the roof, the inspection party should look for natural openings such as scuttle holes and skylights. They should also take note of where the elevator shaft and stairwells come through the roof. It is a good idea to determine the construction of the roof during the inspection. More than one axe blade has been dulled by a firefighter trying to cut a hole in a concrete roof.

The lower floors will provide information that may later prove helpful in cross-ventilation. Note should be taken of the location and types of windows found in the building. Can they be opened easily, or might it be necessary to break them? Can they be broken from the outside if it is desired to use them for ventilation? Plate glass and wire glass windows can prove to be a difficult problem.

7. *Building access.* What problems might arise while trying to gain access to the building during the fire? Are means of entering available on all four sides, or might access be limited to the front and back—or maybe even just the front? Will

Photo 2.2 Fences and similar objects can hamper access to the fire. (Courtesy of Chris Mickal, New Orleans Fire Department Photo Unit)

it be possible to use all natural entry points? Plate glass or rolling steel doors will present problems. Are there gates on the premises that might slow down access to the rear and sides of the building? What length ladders will be required to gain ingress to the building on upper floors?

8. *Hazardous materials.* It is extremely important to determine where any hazardous materials are located on the premises. (See Figure 2.3.) The names of all materials should be written down during the inspection and their hazards determined after returning to quarters after the inspection. An analysis should be made as to how these materials can best be protected from becoming involved during fire operations, and a determination made as to what reactions are likely if they do become involved. Which of the materials are water reactive, heat reactive, poisonous, and so on, and what is the best method of extinguishing a fire involving these materials? Are incompatible materials such as cyanides and acids stored on the premises? It is imperative that plans be made to keep materials such as these from mixing during fire operations. A careless hose line could cause the death of a number of firefighters.

9. *Hazards to firefighters.* In addition to hazardous materials, there might be a number of structural defects that could be a hazard to firefighters during fire operations. The roof is a logical place to start checking for these hazards. When analyzing potential hazards, it is best to consider what condition will most likely exist during the emergency. A television antenna located five feet off the roof does not look hazardous during an inspection, but it could cause a firefighter to lose an eye if the roof is covered with heavy, black smoke at the time of the fire. Evaluate what it will be like when a firefighter comes onto the roof in smoky conditions. This concept should be carried out when evaluating conditions elsewhere in the building. Are there any pits or unprotected vertical openings that might present a hazard to firefighters crawling into the building with a hose line? It is best not to overlook any area of the building when considering this factor.

Figure 2.3 Check for hazards to firefighters

10. *Utility controls.* It might be imperative that the gas, electricity, or water be shut off during the early stages of the emergency. The moment of need is not the time when a search for these controls should be made. The location of the shutoffs and the areas they control should be determined during the pre-fire planning inspection. Additionally, an evaluation of the voltage entering the building should be made to determine if it is feasible to cut the wires. If it is determined that it is practical to do so, then an evaluation of the best place to make the cut should be made.

11. *Salvage.* Thought should be given during the inspection to the salvage problem. Can the floors be expected to hold the water for any length of time? Can the elevator shaft and/or the stairwells be used for water removal? How susceptible is the material stored on the premises to water damage? Will it be necessary to cover the floors to protect them? What are some of the alternative methods that can be used for water removal?

12. *Hose requirements.* How much hose is it going to take to reach various parts of the first floor from the front door? How much will be required if the hose is brought in through other access points? What if lines are advanced up stairwells or fire escapes to upper floors? What lengths will be required if the hose is taken aloft and worked off the dry standpipe system? It is much better to determine the answers to some of these questions prior to the fire than trying to make a snap judgment during the stress of the emergency.

13. *Water supply.* Where are the nearest hydrants located and how much water can be expected from the system? Are there any auxiliary supplies such as water tanks, cisterns, or swimming pools available? If so, how much water can be expected from these sources? Is the water available from these sources adequate to cope with the expected emergency, or might it be necessary to relay water from another source? If so, what source will provide the quickest and best supply?

The Response Size-Up

Many thoughts go through the minds of every firefighter when the bell sounds. Will this one be routine or is this the big one? Is this the one where I might have to crawl in on my hands and knees to save some child? Will people be trapped, which might require the use of special equipment?

The most tangled thoughts, however, will flash through the mind of the officer who is expected to be first in. That individual knows he or she will be responsible for operations until relieved by a superior. That particular officer will have to make the initial determination as to the number of companies needed, where lines are to be placed, and the thousands of little details involving the factors of rescue, extinguishment, exposures, laddering, ingress, ventilation, salvage, and so on. He or she hopes that a pre-fire planning inspection of the building has been made, but more than likely this will not be the case. If this is so, there are a number of factors the officer will have to consider until he or she arrives at the scene where some of the information will then be able to be determined.

The response. Knowing the location of the emergency will provide knowledge as to the number and types of companies that can be expected. If the address is in a residential area, at least two engine companies and a truck company will be received on the first alarm. However, this number can increase sufficiently, depending on the area of response. As many as four engine companies, two truck companies, and additional equipment such as salvage companies, rescue or squad companies, boat companies, and perhaps special equipment might be dispatched if the area has unique problems. The officer who is first on the scene should quickly review in his or her mind the manning of these companies, the types of engine companies that will be responding, whether the truck companies are equipped with aerial ladders or elevated platforms, and the directions from which all companies will be arriving. He or she should also consider the number of additional companies that will arrive on the scene if the fire is beyond the control of the first alarm assignment and it becomes necessary to call for a second or third alarm.

The district. The district in which the emergency is located will have an influence on the size-up. The life hazard in a residential area will normally be considerably less than one in a hotel or apartment house district. The size of the fire will be different in an industrial zone than in a limited commercial area. The speed with which the fire is likely to spread will vary according to the construction in the area and the spacing of the buildings. A response into a brush area will present many different problems than one into a residential area. The factors influencing the area of response should be analyzed as the apparatus moves from the engine house toward the fire.

Photo 2.3 Narrow streets and parked cars hamper response to alarms.

Time of day. Generally the life hazard and the probable size of the fire is much greater between midnight and 4:00 A.M. than it is at 2:00 in the afternoon; however, this may not be the case. The type of occupancy matched with the time of the alarm plays a vital part in the development of a critical situation. The life hazard in a school is much greater at 10:00 in the morning than at 10:00 at night, whereas the life hazard in an apartment house is much greater at 3:00 A.M. than at 3:00 P.M. Thought must be given to the life hazard during the response size-up. Is it likely that people will be in the building or is it more likely that the building will be empty? Will the occupants probably be awake or asleep? Will they be able to get out of the building unassisted or will it be necessary to physically remove most of them?

The time of response will also have an affect on the speed of the response itself. What influence will the time of response have on the traffic problem? A considerable delay can be expected when the work force is leaving a factory or a large building. It can also be expected that crowds will be larger during certain hours than at others. Parked cars may also hamper access to hydrants during some hours, whereas the same hydrants will be freely accessible at other times. All of the factors influenced by the time of the response have to be considered during the size-up.

The weather. The weather will have an impact on the response, the fire spread, and fire tactics. For example, weather conditions such as rain, snow, and fog will have an adverse effect on response time and, in some cases, in locating the fire. Certain areas of the city might become flooded during heavy rains, which could cause a change in response routes. This not only slows down response time but

Photo 2.4 Cold weather can hamper firefighting operations. (Courtesy of Minneapolis, Minnesota, Fire Department)

might bring the company into the fire from a completely different direction than expected by the officer-in-charge of the fire. There have been cases where a company could respond only in one direction from the engine house due to deep flooding in the other directions. Heavy fog will not only slow down response but may make it extremely difficult to locate the fire. Companies have pulled up in front of well-involved buildings during heavy fog conditions before seeing the fire. This reduces the time available for planning hose lays to zero.

The weather may also have a tremendous impact on fire spread. The mere delay in response will give the fire additional time to progress before hose lines can be brought into position to restrict its spread; however, more important is the direct impact it has on contributing to the fire spread. The weather factor most feared is high winds. Winds can lift roof shingles high in the air and drop them on other combustible roofs some distance from the original fire, and they can bend high-pressure hose streams so badly that little of the water reaches the fire. For a number of years, winds in excess of 30 miles per hour have been the leading contributor to conflagrations.

Weather also plays an important part in fire operations. It is much more difficult to attack a fire during periods of heavy snow than it is on warm clear days. It is also different fighting a fire when the outside temperature is 15 degrees than it is when the temperature is 120 degrees. In fact, it is possible that what might have been a greater alarm fire might not even be reported because of the weather. This was illustrated a number of years ago in southern California. A fire occurred in a large, industrial building sometime around midnight. The first report the fire de-

Photo 2.5 Cold weather also has an effect on firefighters. (Courtesy of Minneapolis, Minnesota, Fire Department)

partment received on the fire was the next morning when employees reported to work to find the building burned to the ground. During the night, the area had been blanketed by thick fog. No one had seen the fire, despite the fact that the building was located near a well-traveled street.

The Fireground Size-Up

The fireground size-up can be divided into two distinct phases—the preliminary size-up and the continuous size-up.

The preliminary size-up. The *preliminary* size-up is the immediate estimate of the situation made by the officer-in-charge of the fire upon arrival. (See Figure 2.4.) It is made in accordance to his or her best judgment prior to the commitment of forces and prior to reporting to the dispatch office. It is at this point that all of the knowledge gained by the pre-alarm size-up, the response size-up, and the conditions observed upon arrival are brought into focus. The preliminary size-up forms the basis for the initial deployment of personnel and equipment, and for the calling for additional help if it is determined that it *might* be needed.

The preliminary size-up is the responsibility of the first arriving officer, regardless of rank or seniority. It must be done with initiative and aggressiveness; however, it should not be made in a haphazard or indiscriminate manner. The size-up must be based on adequate knowledge and an intelligent survey of existing condi-

Figure 2.4 Where is it going to go?

tions. There is no place in aggressive firefighting for a first arriving officer who will neglect the responsibility for the size-up and leave this responsibility for the arrival of the chief. It is good practice never to underestimate the size and number of lines needed nor the amount of equipment and personnel that might be necessary. There is every excuse for overestimation, but none for underestimation.

The continuous size-up. The *continuous* size-up, as the name implies, is the continuous and comprehensive estimate of the situation as firefighting operations proceed. Things can change rapidly at a fire. Where it appeared that it might go in one direction, the fire may suddenly reverse itself and proceed in an entirely unexpected direction. Fires might suddenly appear in previously uninvolved portions of the building. Explosions, backdrafts, or flashovers can quickly change what appeared to be a controlled situation into a nightmare. The officer-in-charge must constantly expect the unexpected and be prepared to change tactics to combat it.

THE DIVISION OF FUNCTIONS

The basic firefighting unit of a fire department is a company. A *company* may be defined as a number of people under the command of a company officer assigned to a station with apparatus. The three classifications of personnel assigned to a company are a company commander, one or more apparatus operators, and firefighters.

Theoretically, there are four basic types of companies found in the organization of a fire department. For purpose of explanation, these companies may be referred to as (1) the engine company, (2) the truck company, (3) the salvage company, and (4) the rescue company. (See Figure 2.5.)

It is recognized that very few fire departments today have separate salvage

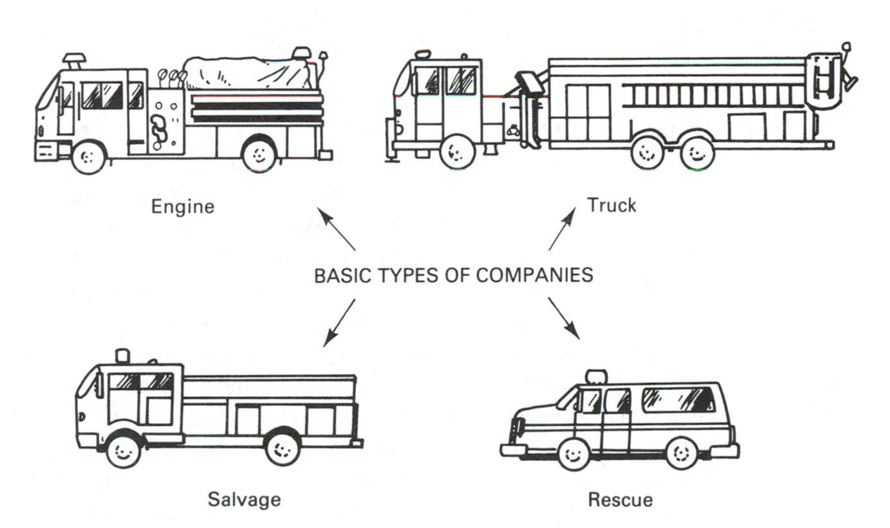

Figure 2.5

companies. In some departments the functions of the salvage company are given to the truck company. In other departments an engine company is assigned salvage company responsibilities at a fire, depending on need. In most fire departments both the engine companies and the truck companies carry some salvage equipment. However, for the purpose of illustrating the importance of salvage operations in the overall coordination of fire operations, it will be considered that a salvage company is a separate entity.

Earlier in this chapter, the functions to be performed at a fire were outlined and the importance of their coordination was discussed. The criticalness of ensuring that the responsibility for performing each of the functions be properly delegated was also evaluated. To meet this need, and for the purpose of organization, coordination, and control, each of the functions are assigned to one of the basic types of companies. Assignments are made on the theory that all companies arrive at the fire at the same time. Of course, this is never the case. It is therefore recognized that, regardless of its basic assignment responsiblities, the first company to arrive at the fire should give first priority to search and rescue. The engine company is generally the first company to arrive.

The engine company is given the responsibility for:

1. Extinguishing the fire.
2. Protecting the exposures.

The truck (ladder) company is given the responsibility for:

1. Laddering operations.
2. Overhaul.
3. Ventilation.
4. Forcible entry.
5. Physical rescue.
6. Controlling the utilities.

The salvage company is given the responsibility for salvage operations.

The rescue company is given the responsibility for emergency medical care.

This is the ideal division of functions and provides for maximum coordination and control. However, it is recognized that most fire departments are not of sufficient size to warrant placing in service and manning all four basic types of companies. Consequently, it is necessary in most departments to combine the functions of two or more of the companies into one. Whenever this is done, the truck company generally assumes the additional responsibility for salvage operations. In some jurisdictions the salvage and emergency medical care functions are assigned to a unit referred to as a *squad company*. Regardless of the manner in which a department is organized, or the titles given to the various companies, it is extremely important that all of the required functions be properly delegated and the members of the companies to which they are given be properly trained in their performance.

Combining functions varies from one department to another, but it must be recognized that responsibility for each of the functions must be properly assigned and adequate personnel must be available to carry out simultaneously the required functions if loss to property and life is to be kept to a minimum. In all respects to the many ways that fire departments are organized, for purpose of illustration and explanation, the required functions will be assigned as outlined under the four basic types of companies.

REVIEW QUESTIONS

1. What is necessary in order to obtain good coordination on the fireground?
2. What is meant by "Football Games Are Not Won on Saturday"?
3. What are the 11 common tasks that must be considered on the fireground?
4. What are a few of the factors considered under existing conditions on the fireground?
5. What is the term applied to the estimate of existing conditions?
6. Who participates in the size-up?
7. When does the size-up commence?
8. For practical purposes, into what three parts can the size-up be divided?
9. When does the pre-alarm size-up manifest itself?
10. When does the pre-alarm size-up generally commence?
11. What is the objective of pre-fire planning?
12. What does pre-fire planning include?
13. How does knowing the size and construction of the building help in pre-fire planning?
14. Within how short a period of time will unprotected steel begin to lose its strength when subjected to severe fire conditions?
15. What factor takes first priority at a fire?
16. What may be the best method of saving life under practical conditions at a fire?
17. What are the three methods by which heat may travel from one part of a building to another?
18. What are the various means by which fire can travel from room to room on the same floor?
19. What are the various means by which fire can travel from floor to floor?
20. What are the various means by which fire can extend from building to building, or from the fire building to material outside of the building?
21. What part does knowledge of the contents of the building play in the pre-fire planning process?
22. What should be learned about the on-site fire protection when making a pre-fire planning inspection of a building?
23. What is an annunciator?
24. When should a building be equipped with a dry standpipe system?
25. Where is it best to start a pre-fire planning inspection for the purpose of evaluating the ventilation problems?
26. What are the natural openings found on the roof that can be used for ventilation?

27. What should be done when hazardous materials are found on the premises during a pre-fire planning inspection?
28. Where is a logical place to start looking for hazards to firefighters when making a pre-fire planning inspection?
29. What utility controls should be found and identified during a pre-fire planning inspection?
30. What should be determined regarding the water supply during a pre-fire planning inspection?
31. What does the district in which a fire is located have to do with the size-up process?
32. What part does the time of day have to do with the size-up?
33. What are the two phases into which the fireground size-up may be divided?
34. What is the preliminary size-up?
35. For what is the preliminary size-up used?
36. Who is responsible for making the preliminary size-up?
37. What is meant by continuous size-up?
38. What is the basic firefighting unit of a fire department?
39. How may a company be defined?
40. What are the three classifications of personnel assigned to a company?
41. What are the four basic types of companies found in the organization of a fire department?
42. What are the two basic responsibilities of an engine company?
43. What are the six basic responsibilities of a truck company?
44. What is the responsibility of a salvage company?
45. What is the basic responsibility of a rescue company?
46. Which company assumes the responsibility for salvage operations if a department does not have separate salvage companies?

CHAPTER 3

Engine Company Operations

OBJECTIVE:

The objective of this chapter is to introduce the reader to the various functions of an engine company. An explanation is offered as to the accepted methods of extinguishing the four classes of fire. An examination is also made of the various methods used to protect both internal and external exposures from the heat given off by a fire. Review questions are provided at the end of the chapter to assist the reader in determining his or her level of comprehension of the chapter contents.

There are more engine companies in fire departments than all other types of companies combined. It is possible to find triple combination engine companies, two-piece companies, manifold companies, tank companies, tanker companies, and more. Engine companies vary by type according to the apparatus assigned to the company. Different types of companies are required in order to cope with the various problems found in different sections of a city or rural area. In some sections of a fire department's response area, fire extinguishment generally requires the use of many small lines, whereas in other sections of the same response area it frequently may be necessary to use heavy stream appliances. Consequently, the personnel, amount of hose, equipment, and pumping capacity required will vary. However, regardless of the manning or equipment found on the company, all engine companies have the same basic responsibilities—to extinguish the fire and to protect the exposures. (See Figure 3.1.)

FIRE EXTINGUISHMENT

The fire triangle introduced in Chapter 1 was used to discuss the theory of fire. It was explained that it is necessary to have all three sides of the fire triangle present in the proper portions in order to have a fire. It follows that once a fire is burning, the removal of one side of the triangle would cause the fire to go out. For many years this simple explanation was accepted; however, there is more to it than that.

Figure 3.1 Engine companies are responsible for extinguishing the fire and protecting the exposures

Photo 3.1 Engine companies require apparatus that carry hose and water and are equipped with a pump. (Courtesy of W. S. Darley and Company)

Shortly after the end of World War II, experiments were conducted using dry chemicals as an extinguishing agent. Much to the surprise of those conducting the experiments, some of the experimental fires were extinguished quicker than expected. In fact, some of the fires were extinguished when it was clear to the experimentalists that none of the sides of the fire triangle had been removed. It was therefore concluded that another element was involved in the extinguishment that was previously unknown. This other element become known as the *chain reaction.* It became apparent that the process of fire involved a chain reaction, and if the chain reaction could be broken, the fire would go out. This theory led to the design of the fire tetrahedron, which added a fourth element to the fire triangle that was previously used to explain the process of fire extinguishment.

The Fire Tetrahedron

The fire tetrahedron is a four-sided figure that incorporates the fire triangle and the added feature of the chain reaction. (See Figure 3.2.) A fire may be extinguished by removing the fuel side, the oxygen side, or the heat side of the tetrahedron or by breaking the chain reaction. The side to be eliminated during extinguishment depends on the type of fire involved.

Classification of Fires

Fires have been classified by type primarily for the purpose of identifying the type of material required for extinguishment. The four accepted classes of fires are:

Photo 3.2 Engine companies are responsible for fire extinguishment. (Courtesy of Chris Mickal, New Orleans Fire Department Photo Unit)

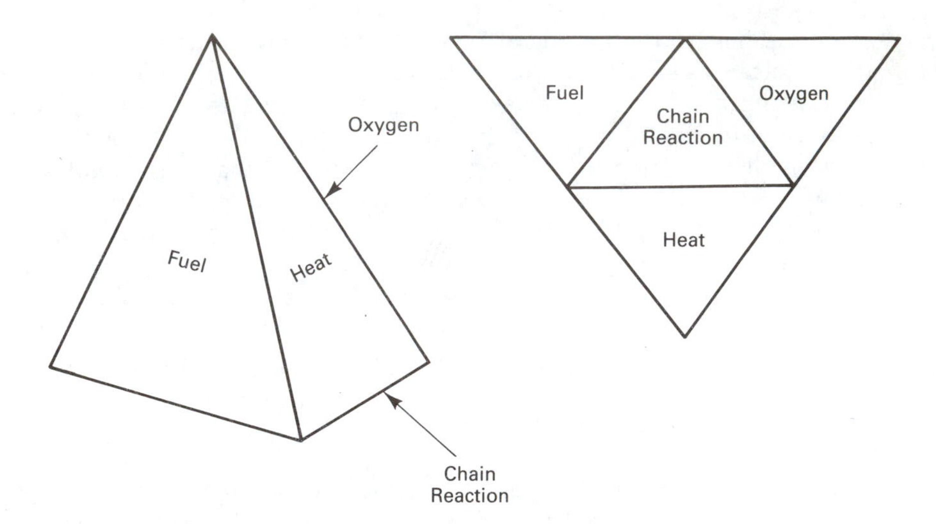

Figure 3.2 The fire tetrahedron

- *Class A Fires* involve ordinary combustible material such as wood, paper, cloth, and so on.
- *Class B Fires* involve flammable liquids, gases, and greases.
- *Class C Fires* involve energized electrical equipment.
- *Class D Fires* involve combustible metals, such as magnesium, titanium, sodium, and potassium.

Extinguishment of Class A Fires

Class A fires involve ordinary combustible material such as wood, paper, cloth, and the like. (See Figure 3.3.) During the burning process of Class A materials, a certain amount of heat is released, depending on the type of material involved. For example, in the process of complete combustion one pound of wood will give off approximately 8,000 to 9,000 Btus, and one pound of paper will give off approximately 7,500 Btus. A Btu (British thermal unit) is a heat quantity of measurement and is normally defined as the quantity of heat required to raise the temperature of one pound of water one degree Fahrenheit. The rate at which the heat is released is directly related to the intensity of the fire.

Some of the heat released during the burning process is absorbed by the air or surrounding material; however, for purpose of extinguishment it should be considered that it is necessary for the extinguishing agent to absorb heat faster than it is being generated. The final objective is to reduce the temperature of the burning material and the surrounding atmosphere to below the ignition temperature of the burning material.

Water. The majority of all fires encountered by fire departments are of the Class A type. Class A fires are generally extinguished by removing the heat side of

Figure 3.3 Class A fires are fires in ordinary combustible material

the fire tetrahedron. The material most commonly used for removing heat is water. Water is most often used, not only because of its availability but also because of the characteristics of its composition.

The ability of an extinguishing agent to absorb heat is referred to as its *specific heat, thermal heat,* or *heat capacity,* the terms being synonymous. The specific heat of an agent is expressed by the number of Btus absorbed by one pound of the agent as its temperature is raised one degree Fahrenheit. Water has a specific heat of 1.0, which is higher than most other materials. For instance, ice has a specific heat of approximately 0.5, only half that of water. There is no significant difference between the specific heat of fresh water and that of salt water.

To place the numbers in perspective, 100 Btus will be absorbed by 100 pounds of water as the temperature of the water is increased one degree Fahrenheit. Since water weighs approximately 8.33 pounds per gallon, 8.33 Btus will be absorbed by one gallon of water as the temperature of the water is increased one degree Fahrenheit. Looking at it another way, if the temperature of one gallon of water is raised from 60 degrees Fahrenheit to 212 degrees Fahrenheit during the extinguishment process, it will absorb approximately 1,266 Btus (8.33 × 152). Unfortunately, in most cases the temperature of the water is normally not raised to 212 degrees and a good portion of the water applied will not display any significant rise in temperature. In fact, it is estimated that approximately 90 percent of the water applied by straight streams is involved in runoff, which means that the water has only utilized a maximum of 10 percent of its absorption capabilities.

In order for water to absorb heat directly from the burning material, there must be surface contact between the burning portion of the material and the water. Consequently, the greater the surface contact of a given amount of water with the burning material, the greater the absorption capability. It follows that spray streams and fog streams have the capability of absorbing more heat than straight streams; however, it should not be concluded that these streams are superior for firefighting under all conditions. It must be remembered that it is necessary for the water to reach the burning material before heat absorption commences, and at times the reach of the stream and the penetration ability of the stream will take priority over the size of the droplets.

Although the heat absorption capability of water is extremely effective when compared with other materials, it is even more effective as the water changes from a liquid to a gas. The absorption that takes place during this process is referred to as the *latent heat of vaporization.* The latent heat of vaporization is defined as the amount of heat absorbed or given off as a substance passes between the liquid and gaseous phases. The latent heat of vaporization of water at its boiling point (212° F.) is 970.3 Btus per pound. This means that 970.3 Btus will be absorbed by water as it changes from a liquid to steam. Compare this with the 152 Btus that are absorbed as one pound of water is raised from 60° F. to 212° F. The result is that approximately 6.4 times more heat is absorbed as water is changed from a liquid to steam than when the temperature of the water is raised from 60° F. to 212° F.

It should be noted that heat is absorbed as water is changed from a liquid to steam; however, the temperature of the liquid does not increase. The absorption of the additional heat reduces the volume of the liquid, with volume reduction continuing until the last drop of water has been converted to steam.

Combining the effectiveness of water due to its specific heat and its latent heat of vaporization, it can be seen that if one pound of water is raised from 60° F. to 212° F. and then completely vaporized into steam, it would absorb approximately 1122.3 Btus (152 + 970.3). Since water weighs approximately 8.33 pounds per gallon, it is possible to absorb approximately 9,349 (8.33 × 1,122.3) Btus per gallon during this complete process. Ironically, this is about the same amount of heat that is given off by one pound of wood during the process of complete combustion. Absorption of 9,349 Btus from a gallon of water is theoretically possible, but it should be kept in mind that it is seldom achieved during firefighting operations.

In addition to the heat absorption capability of water as it is changed to steam, there is an additional value to be gained that aids in extinguishment once the change has been completed. This is due to the tremendous expansion that takes place. The expansion ratio is approximately 1,700 to 1 at a normal atmospheric pressure of 14.7 psi and a temperature of 212° F. However, the expansion ratio is a function of temperature. The ratio is approximately 2,400 to 1 when the fire area is 500° F. and approximately 4,200 to 1 when the temperature is 1,200° F. This expansion forces smoky and noxious gases from the involved structure and reduces the amount of oxygen available to support combustion. It should be noted, however, that although the primary value of the steam expansion is the purging of the air and the resultant reduction of the available oxygen needed to support combustion, fires in Class A materials are normally extinguished by the absorption of heat, not by the smothering effect created by the steam. The smothering effect has the tendency to suppress flaming, but it is the cooling effect that extinguishes the fire.

Water additives. Several chemicals are added to water in an attempt to increase its effectiveness as an extinguishing agent. Materials have been added to lower the freezing temperature, to thicken the water, to reduce its surface tension, and to reduce its friction loss. The two most common products in use from a fire-fighting standpoint are known as wet water and slippery water.

Wet Water. Wet water is water to which an additive has been introduced in order to reduce the surface tension of the water. Plain water has a relatively high surface tension, which reduces its effectiveness in deep-seated fires such as those found in couches, mattresses, bales of cotton, and stacked hay.

The effect of the high-surface tension of water can be observed when water is poured from a glass. (See Figure 3.4.) Some of the water will stick to the inside of the glass in the form of droplets. When water is applied to deep-seated fires this surface tension reduces the ability of the water to penetrate. The result is that the fire is extinguished on the surface but continues to burn beneath the surface. Wet water, on the other hand, will penetrate beneath the surface and extinguish deep-seated and hidden fires. Wet water is not only effective for use in these types of fires but it also has proven to be effective on ordinary combustibles as it will penetrate the subsurface and assist in the prevention of rekindles.

Wet water may be produced by premixing the additives with the water prior to the fire or by using proportioning equipment when the demand for its use arises on the fireground.

Slippery Water. Friction loss in hoses caused by water moving through the hose is an age-old problem in the fire service. The problem increases in intensity as greater amounts of water are required from a hose stream. When large amounts of water were required from hose streams, the general practice for many years was to increase the number or size of lines feeding the nozzles. This practice prevailed as the general opinion among fire officials was that little could be done to reduce the friction loss in the hose.

In 1948, experiments were conducted with the use of polymers as an additive to water. It was found that the polymers not only reduced the friction loss in the hose but also increased the amount of water that could be moved through a hose line. One experiment indicated that with the use of polymer additives it was possible to get as much water from a 1½-inch line as it was from a 2½-inch line using plain

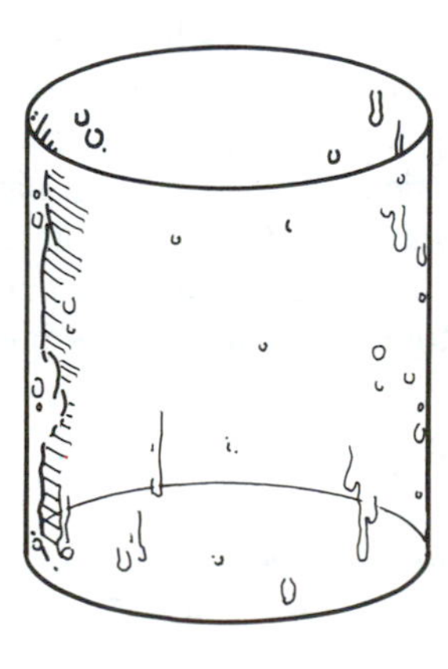

Figure 3.4 Water has a high surface tension

water, and that a 2½-inch line was able to deliver as much water as a 3-inch line. Additionally, tests indicated that it is possible to increase the nozzle pressure and the reach of a stream with the use of polymer additives.

The product resulting from the addition of polymers to water became known as slippery water. Slippery water can be produced with salt water as well as fresh water. It is nontoxic and can be used with all firefighting equipment; however, special care should be taken when it is used on small hand-held lines. Although it is possible to obtain as much water through a 1½-inch line when slippery water is employed as through a 2½-inch line using ordinary water, the slippery water does not reduce the nozzle reaction. Consequently, the same care should be observed as if a 2½-inch line was placed into operation.

Dry chemicals. Multipurpose dry chemical extinguishers can be effectively used to extinguish small fires in Class A material. Extinguishment takes place by breaking the chain reaction and by the decomposition of the agent, which leaves a sticky residue on the burning material. The sticky residue in effect assists in the extinguishing process by eliminating the oxygen side of the fire tetrahedron through the formation of a seal between the burning material and the air. The seal not only extinguishes the fire but also assists in the prevention of reignition. Dry chemicals, however, have a limited cooling capability; consequently, they do not extinguish deep-seated fires. Water should be used as a followup when fires of this type are encountered.

Multipurpose fire extinguishers also have the capability of extinguishing Class B and Class C fires. Additional discussion on these extinguishers is included in the next section.

High-expansion foam. Foam consists of masses of air or carbon dioxide bubbles dispersed in water. The air or carbon dioxide bubbles are entrapped in the water by the addition of stabilizing agents that are introduced at the time the foam is being developed. The foam will expand during the development phase, the amount depending on the materials used for forming the foam. Low-expansion foams have an expansion ratio up to about 20 to 1, whereas high-expansion foams have an expansion ratio as high as 1,000 to 1.

High-expansion foams are used to fill an enclosed area completely. Extinguishment takes place by a dual process. The foam carries water to the fire and provides a cooling effect, and it replaces the air, which helps in the extinguishing process by blanketing the fire. It is particularly effective on basement or cellar fires but is also useful on fires in holds of ships or in engine rooms. Whenever foam is used on this type of fire, every effort should be made to maintain the area where the fire is located in as much an air-tight compartment as possible.

Extinguishment of Class B Fires

Class B fires involve flammable liquids, gases, and greases. (See Figure 3.5.) Class B fires are generally extinguished by removing the oxygen side of the fire tetrahe-

Photo 3.3 High-expansion foam can be used effectively on several types of fire. (Courtesy of Chicago Fire Department)

dron or breaking the chain reaction. There are several extinguishing agents available that have the capability of achieving one or both of these objectives.

Dry chemicals. Dry chemicals are powders that are formed by grinding dry crystals and adding substances that will cause the powder to flow easily and resist moisture and caking. The developed product is nontoxic; however, breathing too much of the material can cause irritation to the throat and lungs.

Dry chemicals are extremely effective on Class B fires. When the material is projected into the fire area, the fire goes out almost immediately. Although it is known that smothering and radiation shielding contribute to the extinguishment of the fire, the quickness with which flames are eliminated suggests that the principal factor in extinguishment is the breaking of the chain reaction.

When dry chemical extinguishers are used on a Class B fire, discharge should be commenced some distance from the fire. The powder is initially released under considerable force. If it discharged too close to the burning material it may cause the fire to spread before it can be brought under control.

It should be remembered that dry chemicals have a relatively limited cooling capability; consequently, the extinguisher operator should be on the alert for possible reignition if the Class B material is burning in a metal container. Reignition is the result of the container being hotter than the ignition temperature of the material involved. Reignition is much more likely if burning has been taking place for any length of time.

There are two basic types of dry chemical extinguishers available. The multipurpose chemical extinguisher that was previously discussed is effective on Class A,

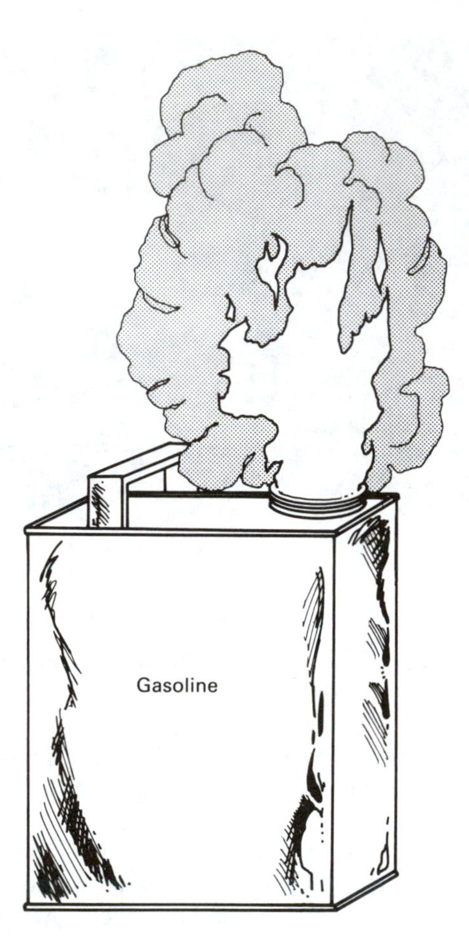

Figure 3.5 Class B fires are those in flammable liquids, gases, and greases

Class B, and Class C fires. Regular or ordinary dry chemicals are effective on Class B and Class C fires.

The residue from a dry chemical extinguisher is a white powder that will cover everything it touches. The residue from a multipurpose extinguisher should be thoroughly removed from all metal parts as quickly as possible after the fire has been extinguished. Failure to to so will cause corrosion of some metals. If metals are not involved, the residue can be washed away with water or possibly removed by the use of a vacuum cleaner.

It is good public relations for the fire department to remove the residue. In some instances this cannot be done because it may require disassembling the equipment in order to remove the residue adequately. In such cases, the owner or manager should be advised of the decontamination needed.

Carbon dioxide. Carbon dioxide is a nontoxic, noncorrosive, nonconductive, colorless, tasteless, odorless gas. When used as an extinguishing agent it is normally stored under 800 to 900 psi pressure, which causes it to be converted to a liquid. The pressure under which it is stored provides the power for its discharge.

Carbon dioxide is effective on both Class B and Class C fires. It was the primary extinguishing agent used during World War II for flammable liquid fires. Its use on this type of fire, however, has given way to dry chemicals due to their more effective extinguishing capabilities.

Carbon dioxide is discharged as a large white cloud that contains particles of dry ice. It is effective as an extinguishing agent primarily because of its smothering effect, which displaces the oxygen content of the atmosphere and therefore removes the oxygen side of the fire tetrahedron. It has a vapor density of 1.5, which causes it to settle over the burning material when discharged above the fire.

Carbon dioxide should be discharged above and directly on the fire. The range of the discharge is extremely limited, which means that a close approach to the fire is mandatory if the fire is to be extinguished. When carbon dioxide is discharged, the "snow" it produces provides some degree of cooling action, but the cooling effect is very limited. Once the fire is extinguished, the same caution should be taken as with the use of dry chemicals when Class B fires are encountered in metal containers. The operator must be alert for a possible reflash as the carbon dioxide cloud is dispersed.

Foam. As previously mentioned, foam consists of air or carbon dioxide bubbles dispersed in water. High-expansion foams are used on Class A fires, and low-expansion foams are used on Class B fires. Low-expansion foams are those having an expansion ratio of 20 (or less) to 1. The two general types in common use are chemical foams and mechanical foams. Although the two are formed by the use of different materials, their use as a firefighting agent is the same.

Chemical Foam. Chemical foam is a thick, stable mass formed by a chemical action with water, which when used with a foaming agent entraps carbon dioxide gas in the bubbles. This type of foam takes time to set up, and the chemicals used in its composition are corrosive. The powder used to form the foam has a tendency to cake.

Mechanical Foam. Mechanical foam is a thick, stable mass that contains air bubbles. It is formed by mixing water with a protein-based liquid that is made from animal products.

Use of Foam. Foam is effective on Class B fires in containers as well as Class B spill fires. It extinguishes the fire by covering the fire and blanketing it from the atmosphere. Once extinguished, the fire is relatively safe from a reflash because the blanket of foam will exclude the air for some time—depending on the thickness of the foam blanket and the composition of the foam.

In open-container Class B fires, foam should be applied by projecting the foam stream to the *far side* of the container and allowing it to cascade over the burning liquid. Spill fires in the open should be attacked by directing the foam stream on the surface in *front* of the fire, which will push the foam blanket over the fire. The foam stream should not be directed *into* the fire, whether the fire is in a closed container or as a result of a spill. To do so would most likely result in the fire spreading. (See Figure 3.6.)

Figure 3.6

Foam can also be used to prevent Class B fires under certain circumstances. It can be extremely effective in blanketing a large spill to prevent a fire and eliminate a hazardous condition. It can also be used to blanket a runway in anticipation of a fire in an aircraft that is forced to land with its wheels up; however, its usefulness for this purpose is limited because of the time involved.

Alcohol Foams. Regular foams are not effective on all types of Class B fires. For example, they are not effective on alcohol fires. The alcohol breaks down the foam before it can extinguish the fire; however, special foams known as *alcohol foams* are available for use on these fires. The method of application for these foams is the same as for regular foams. It should be noted that alcohol is used in many cases to boost the octane rating of gasoline. In these cases, gasoline fires may require the use of an alcohol foam.

Light Water. Light water (also known as *Aqueous Film Forming Foam—AFFF*) is synthetically developed and contains fluorocarbon surfactants that cause a thin aqueous film to float on top of fuels. It is a highly versatile product that is three to five times more effective than ordinary foams on Class B fires. One type available is effective on regular Class B fires, such as gasoline and oil, as well as alcohol fires. It is also effective on deep-seated Class A fires. Some fire officials have stated that light water appears to have the knock-down characteristics of dry chemicals and the holding power of foam.

Halogenated extinguishing agents. Halogenated extinguishing agents are available for use by firefighters in portable fire extinguishers. These agents are formed by replacing one or more hydrogen atoms in hydrocarbons with atoms from the halogen series. The products involved in the halogen series are fluorine, chlorine, bromine, and iodine.

Halogenated extinguishing agents have been available for many years, but some of the earlier agents, such as carbon tetrachloride, were excessively toxic upon thermal decomposition and were taken out of service as fire extinguishing agents. The two common agents in use at the present time are Halon 1301 and Halon 1211. Neither of these has demonstrated any adverse health effects during the extinguishment process when used as intended. It is possible, however, for firefighters to experience some dizziness if exposed to too high a concentration of halogenated vapors for any length of time. Consequently, it is good practice to require the use of breathing apparatus in all cases where firefighters may be exposed to vapors from any type of halogenated extinguisher agent.

Halon 1301 and 1211 are both gases. They are stored as a liquid due to their storage under high pressure. The stream from the extinguisher is discharged as a liquid but rapidly changes to a gas.

Use of these agents has demonstrated a complete and rapid extinguishment of fire. The actual cause of extinguishment is not presently known, but all indications are that extinguishment takes place by breaking the chain reaction of the fire. An additional advantage of the use of these agents for extinguishment is that they leave no corrosive or abrasive residue.

Extinguishment of Class C Fires

Class C fires involve energized electrical equipment. (See Figure 3.7.) Once the equipment is deenergized, the fire becomes a Class A or Class B fire and can be handled accordingly. The materials normally used for extinguishing these fires are nonconductive in nature in order to protect the user from severe electrical shock.

There is no available extinguishing agent for use on Class C fires only. All the agents used on Class C fires are also effective on either Class A fires, Class B fires, or both. The nonconductive agents most commonly used are carbon dioxide, dry

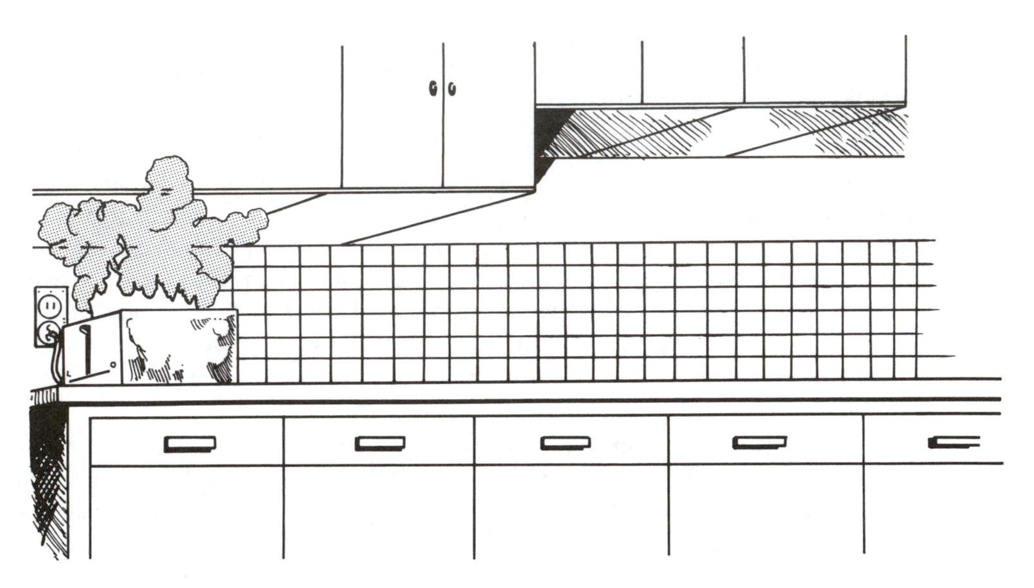

Figure 3.7 Class C fires are those in energized electrical equipment

chemicals, and the halon extinguishers. Although water is conductive, it can be used by experienced firefighters in certain circumstances.

For firefighting purposes, energized electrical equipment can be broken down into two categories: regular equipment and sensitive equipment. Examples of sensitive equipment are computers and equipment containing large numbers of relays. All of the listed extinguishing agents can be used on regular equipment; however, extinguishing agents for use on sensitive equipment should be limited to carbon dioxide and the halons. These agents do not leave a residue, which would extend the damage to the equipment. If both carbon dioxide and a halon are available, carbon dioxide is generally the preferred agent.

Use of water. Water may be used to extinguish electrical fires if it is used properly and with due precaution; however, it should be used only as a last resort. Every attempt should be made to deenergize the equipment prior to applying water. In most cases it is safe to let the fire burn while attempting to deenergize the circuit. The equipment is normally already destroyed and it is only a matter of extinguishing the fire. Water can be used to protect the exposures while the circuit is being deenergized. However, if it is necessary to attack the fire with water, then certain precautions should be observed. It should be noted that although it is possible for a nozzleman to receive a small degree of shock, there is no recorded incident where a firefighter has been killed by the use of a hose stream on an electrical fire.

Where possible, water in the form of fog should be used on electrical fires. The air space between the water droplets limits the conductivity and makes the use of the water relatively safe. Care should be taken not to work too close to the electrical source, however. In fact, extinguishment should be made as far from the electrical source as possible. Caution should also be taken for the nozzleman not to stand in a puddle of water, and not to let a puddle of water form around him or her during the extinguishment process. Severe electrical shock could develop if this precaution is not observed.

It is also possible to use straight streams on electrical fires. As a precautionary measure, it is best to use them only where the fire is beyond the range of fog or spray streams, and to apply the water at a distance that approximates the effective range of the stream.

Extinguishment of Class D Fires

Class D fires involve combustible metals. (See Figure 3.8.) Note that the classification does not refer to metal fires but to fires in combustible metals. Most metals will burn if sufficient heat is applied; however, fires in ordinary metals such as iron, steel, aluminum, and copper can be extinguished by cooling with water. The metals referred to in the classification of combustible metals are those metals that require special extinguishing agents or special application of water in order to extinguish. These metals are magnesium, sodium, potassium, uranium, titanium, lithium, thorium, hafnium, zirconium, and plutonium.

Extinguishing agents for these fires can be found in extinguishers and in cans located on the premises of some industrial plants where these metals are processed, used, or stored. Unfortunately, no single extinguisher or material is effective on all the different types of metals. Equally unfortunate is the fact that the only extinguishing agent found on firefighting apparatus that can be used on any of the metals is water. The common extinguishing agents found in extinguishers on apparatus are not suitable for these fires and can result in dangerous or toxic reactions if used.

Of all the metals mentioned, the one most likely to be encountered by firefighters is magnesium. It may be found in finely divided form, in chips or chunks, or in finished products. Although the general principle is not to use water on magnesium fires, water can be used if proper methods are utilized and the reactions expected are understood.

When magnesium burns, it burns as a molten mass. When water is applied, the water penetrates the surface of the burning metal and expands. Water has an expansion ratio of 1,700 (or greater) to 1. This means that a single drop of water after penetrating the surface of magnesium will expand to at least 1,700 times its original size. The result is an explosion that throws burning magnesium particles in all directions. These particles often burn through a firefighter's protective clothing. Firefighters should expect violent reactions when straight streams are applied to burning magnesium. Eye goggles should always be worn.

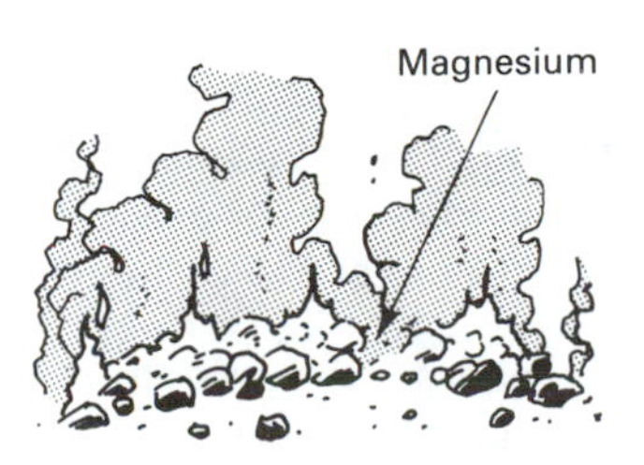

Figure 3.8 Class D fires are those in combustible metals

The success for using water on magnesium fires depends on how quickly a large amount of water can be applied to the fire. It is best when attacking small fires to use fog or spray streams. Fog laid gently on the surface of the magnesium will restrict penetration, and, although popping will occur, the overall effect will be that the magnesium will be cooled and extinguished. Regardless, as a precautionary measure it is a good idea to operate from as great a distance from the fire as possible.

Large fires in magnesium present a much greater problem. It is impossible to extinguish the fire without a great number of violent reactions occurring. Firefighters should operate as far from the fire as possible within the range of the hose streams. The initial attack should be made with large quantities of water. It may be necessary and wise to use heavy streams. If an insufficient amount of water is applied, it will only intensify the fire and possibly spread it over a wide area. When adequate quantities of water are used, after the initial explosion and fireworks, the magnesium should be cooled until extinguishment is achieved.

It is good practice not to use water on any of the other so-called combustible metals. It is also good practice always to wear protective clothing, including breathing apparatus, when working on combustible metal fires, regardless of their size.

PROTECTION OF EXPOSURES

The second function for which an engine company is responsible is the protection of exposures. Although there are a number of methods used for protecting exposures, the most effective method at a majority of fires is to make an aggressive attack on the fire. This chapter will discuss exposures in general. Tactics used for their protection are included in Chapter 6.

An *exposure* may be defined as anything in close proximity to the fire that is not burning but that might start burning if some type of corrective action is not taken quickly. As a review of Chapter 1, it should be remembered that if sufficient oxygen is available, a burnable material will start burning once its temperature is raised to its ignition temperature. As a further review, heat can travel from a burning material to a nonburning material by direct contact, radiation, conduction, or convection. (See Figure 3.9.) Consequently, to protect an exposure it is necessary to protect it against all four sources that are capable of raising it to its ignition temperature. An exposure may be found a few feet or several hundred feet from the main body of the fire, depending on the size and type of fire. They are divided into two general types: interior exposures and external exposures.

Internal Exposures

Internal exposures are ones located inside a building. They may be found relatively close to the main body of fire, in the same room as the main body of fire, in adjacent rooms on the fire floor, or on floors above or below the fire floor.

Direct contact. In most cases the protection of exposures from direct contact is made by an aggressive attack on the fire together with the cooling of the

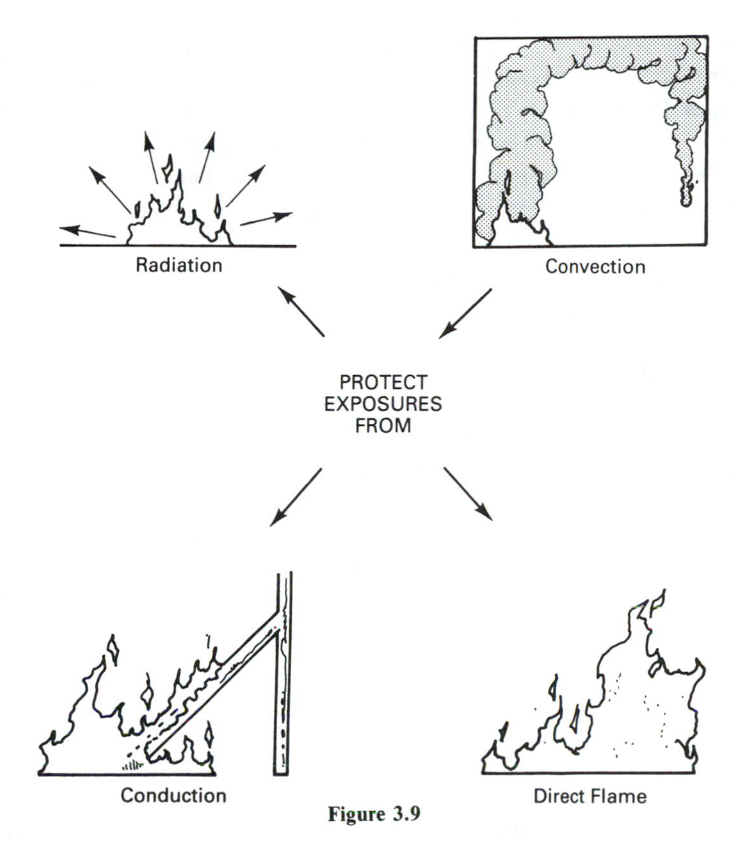

Figure 3.9

exposures by the use of water. An example is a room partially involved with fire. Somewhere in the room the fire is making direct contact with an exposure. Normal procedure is to enter the room and knock down the fire as rapidly as possible. The exposure problem is therefore eliminated, providing, of course, that the fire has not spread to other areas by conduction, convection, or radiation.

It is possible in some small fires for an exposure that is in direct contact with the fire to be protected by moving it away from the fire. This type of protection of exposures from direct contact, however, is generally limited to small fires.

Radiation. Other than by an aggressive attack on the fire, interior exposures are generally protected from radiation by cooling, moving the object, or closing openings to other rooms or areas. Protection of interior exposures by cooling is normally accomplished during the general attack on the fire.

An example of protecting interior exposures from radiation is a burning piece of furniture that is threatening other furniture around it. It is generally a simple matter of moving the other furniture away from the fire until a line can be brought in to extinguish the fire.

Exposures in rooms adjacent to a fire area can be protected from the radiated heat by closing a door. This method is more often used by occupants of a building or by built-in fire protection than it is by fire personnel. An occupant who discovers a fire in a room, closes the door, and then runs to call the fire department has protected the exposures outside of the room from radiation. The closing of a fire door by the melting of a fusible link accomplishes the same thing. Although this type of protection is normally provided by built-in devices or actions of the occupants, it should not be ignored by fire personnel if it can prove to be effective at a fire.

Conduction. Exposures are generally protected from conducted heat by moving them out of contact with the heated body. This type of protection of exposures occurs more often on ships than it does in buildings. An example is a fire in the hold of a ship. Material in the adjacent holds is protected by moving it away from the metal bulkheads. The same procedure may occur with a fire in a stateroom. In this case it may be necessary to move objects away from the four adjacent bulkheads surrounding the room and from the decking above and below the stateroom.

It is usually difficult to protect interior exposures from conducted heat because the resultant fires generally occur in concealed areas. A partition fire is a good example. It should be kept in mind, however, that exposures are ignited by conduction and a thorough check should be made of all areas where extension of the fire by this source is possible.

Convection. Protection of interior exposures from convection presents the biggest problem. Exposures may be ignited in the fire area, in adjacent rooms, and sometimes many floors above the fire floor by convected heat. The truck company generally plays the biggest part in protecting exposures from this source of heat by controlling the flow of the heat through ventilation. In multiple-story buildings, however, it is necessary for the engine company to take positive action early to prevent the spread of fire to upper floors. This means that the engine company must gain control of all vertical openings. This requires extending lines to the floor above the fire and using hose streams to protect the upper floors from the convected heat. Tactics used will be discussed in Chapter 6, but at this point it is worth mentioning that these tactics are risky and should be carried out only with the primary consideration of safety to personnel.

External Exposures

External exposures are ones that are outside the fire building, such as another building, a vehicle, a tree, and so on. They may be located relatively close to the main body of fire or in some cases a considerable distance away.

Direct contact. Protection of an exposure from direct contact with the fire generally manifests itself when a building or a large part of a building is well involved with fire and exposures are relatively close to the fire building. When this occurs it is generally best to cool down the exposures with a generous amount of water followed by an aggressive attack on the fire.

Radiation. Exposures threatened by radiated heat from the fire can be classified as moveable or nonmoveable. Nonmoveable exposures are buildings, trees, and the like, whereas moveable exposures are vehicles, boats, or other objects that are not stationary. Nonmoveable exposures are generally protected by wetting down or setting up a water curtain, whereas moveable exposures can be wetted down or moved. When lines are laid to a fire having exposures that are in immediate danger of igniting, the first lines should normally be used to protect the exposures and water should not be projected onto the fire until all the exposures are adequately protected. Where possible, those exposures capable of being moved (cars, buses, boats, etc.) should be moved. Although moveable exposures can be protected by wetting them down, it is better to move them. The exposure problem is eliminated once they are moved, and the energy and lines required to keep them wetted can be used elsewhere.

It is possible for radiated heat to enter exposed buildings through openings such as windows. This hazard can be eliminated by sending firefighters into the exposed building to close the openings.

On large fires or on some multiple-story fires it may be necessary to set up water curtains to protect the exposures from radiated heat. It may even be necessary to use heavy stream appliances for this operation. In some cases it is possible to protect the exposure by setting up a water curtain from inside the exposure. Lines can be taken inside the exposed building and the water curtain set up by using spray streams out the windows of the exposed building.

Conduction. Protection of exterior exposures from conducted heat is normally not a problem; however, it may manifest itself in large fires that are throwing burning embers high into the air. These embers present an exposure hazard when they land on combustible roofs. This type of problem is generally present at wildland fires that are burning in close proximity to built-up areas. The problem will exist downwind. Protection of exposures from this cause is provided by having engine companies patrol downwind from the fire.

Convection. Protection of exposures from convected heat is normally not a problem if the exposures are limited to one or two stories in height. In can become a problem, however, if the exposures are multiple-story buildings. The methods used to protect the exposures from this heat source is the same as that for radiated heat.

REVIEW QUESTIONS

1. What are some of the various types of engine companies found in departments?
2. What is it that identifies one type of engine company from another?
3. Why are various types of engine companies needed?
4. In discussing the theory of fire, for what purpose is the fire triangle used?
5. What are the four sides of a fire tetrahedron?
6. What is the primary purpose for which fires have been classified by type?

7. What is a Class A fire?
8. What is a Class B fire?
9. What is a Class C fire?
10. What is a Class D fire?
11. What is a Btu?
12. How many Btus will one pound of wood give off during the process of complete combustion?
13. How many Btus will one pound of paper give off during the process of complete combustion?
14. What is the final objective when applying a heat-absorbent material to a fire?
15. What type are the majority of all fires encountered by fire departments?
16. How are Class A fires generally extinguished?
17. What is meant by the specific heat of a material?
18. How is the specific heat of an agent expressed?
19. What is the specific heat of water?
20. What is the difference between the specific heat of fresh water and that of salt water?
21. How many Btus will be absorbed by one gallon of water as the temperature of the water is increased one degree Fahrenheit?
22. How many Btus will be absorbed by one gallon of water if the temperature of the water is raised from 60 degrees Fahrenheit to 212 degrees Fahrenheit during the extinguishment process?
23. What percent of the water applied by straight streams during a fire is involved in runoff?
24. What is necessary in order for water to absorb heat directly from the burning material during a fire?
25. What is the definition of the "latent heat of vaporization"?
26. What is the latent heat of vaporization of water at its boiling point (212° F.)?
27. How much more heat is absorbed by water during the latent heat of vaporization than by raising water from 60° F. to 212° F.?
28. How many Btus will be absorbed by water if it is raised from 60° F. to 212° F. and then completely vaporized into steam?
29. What is the expansion ratio of water at a normal atmospheric pressure of 14.7 psi?
30. What part does the expansion ratio of water play in fire extinguishment?
31. For what purposes have materials been added to water in an attempt to increase its effectiveness as an extinguishing agent?
32. What are the two most common products derived from adding chemicals to water to increase its effectiveness as an extinguishing agent?
33. What is wet water?
34. How high is the surface tension of plain water?
35. What is slippery water?
36. What is added to water to obtain slippery water?
37. How much water can be obtained from a 1½-inch line when slippery water is added into the hose stream?
38. What effect does slippery water have on the nozzle pressure and the reach of a stream?
39. What effect does slippery water have on the nozzle reaction?

40. What is the primary method by which dry chemicals extinguish a Class B fire?
41. What is the cooling capacity of dry chemicals?
42. What does foam consist of?
43. What is the expansion ratio of low-expansion foam?
44. What is the expansion ratio of high-expansion foam?
45. For what are high-expansion foams used?
46. What is the dual process that takes place when high-expansion foam is used for extinguishment?
47. Where is high-expansion foam used most effectively?
48. By what method are Class B fires normally extinguished?
49. What are dry chemicals?
50. What is the toxic effect of dry chemicals?
51. What is considered the primary method by which dry chemicals extinguish Class B fires?
52. What is the effect if dry chemicals are discharged close to the burning material?
53. What are the two different types of dry chemical extinguishers available?
54. Which of the dry chemical extinguishers will cause corrosion of metal parts if the powder is not removed immediately after the fire is extinguished?
55. What are the characteristics of carbon dioxide?
56. Under what pressure is carbon dioxide normally stored in an extinguisher?
57. What is the primary method by which carbon dioxide extinguishes a Class B fire?
58. Where should carbon dioxide be discharged in relationship to the fire?
59. What are the two general types of low-expansion foams in common use?
60. What is chemical foam?
61. What is mechanical foam?
62. How does foam extinguish a Class B fire?
63. How should foam be applied to a fire in an open container?
64. How should foam be applied to a spill fire of flammable liquid?
65. What are alcohol foams?
66. What is light water?
67. How much more effective is light water than ordinary foams on Class B fires?
68. Is light water effective on alcohol fires?
69. What are halogenated extinguishing agents?
70. What are the two common halogenated extinguishing agents in common use at the present time?
71. Are the fumes from Halon 1301 and 1211 toxic?
72. What is the actual method by which halon extinguishers eliminate a fire?
73. What are the characteristics of those agents normally used for the extinguishment of Class C fires?
74. What agent is effective for use on Class C fires only?
75. What are the nonconductive agents most commonly used on Class C fires?
76. What are the two general categories of energized electrical equipment?
77. Which of the nonconductive agents used on Class C fires can be used for regular energized equipment?

78. Which of the nonconductive agents used on Class C fires can be used for fires in sensitive electrical equipment?
79. How many firefighters have been killed as a result of using a hose stream on an electrical fire?
80. Where possible, what form of hose stream should be used on electrical fires?
81. What extinguishing agent should be used on ordinary metal fires such as iron, steel, aluminum, or copper?
82. Which burning metals require the use of special extinguishing agents?
83. Of all the combustible metals, fires in which type of metal are the most likely to be encountered by firefighters?
84. Upon what does the success of using water on magnesium fires depend?
85. What is the definition of an exposure?
86. What are the two general types of exposures?
87. What is an internal exposure?
88. How are internal exposures protected from direct contact with the fire?
89. How are internal exposures protected from radiated heat?
90. How are internal exposures protected from conducted heat?
91. How are internal exposures protected from convected heat?
92. What is an external exposure?
93. How are external exposures protected from direct contact with the fire?
94. How are external exposures protected from radiated heat?
95. How are external exposures protected from conducted heat?
96. How are external exposures protected from convected heat?

CHAPTER 4

Truck Company Operations

OBJECTIVE

The objective of this chapter is to introduce the reader to the various functions of the truck company. Rescue operations (including search procedures), laddering operations, forcible entry operations, ventilation operations, and overhaul operations are explored. Review questions are provided at the end of the chapter to assist the reader in determining his or her level of comprehension of the chapter contents.

In some parts of the country truck companies are referred to as ladder companies. Regardless of the terminology, the functions performed by these companies are the same. It might be noted that all fire departments do not have truck companies. In these instances, the truck company's functions might be carried out by the second or third arriving engine company or the squad company. This might also be the case in some rural areas that do not receive a truck company on the first alarm response due to the extreme distances involved.

Members of truck companies do not normally become involved in actual fire extinguishment operations. However, the functions performed by members of these companies are extremely important to the successful, rapid control and extinguishment of a fire, to the safety of the firefighters at the emergency, and to keeping the overall fire loss to a minimum. If they carry out their tasks effectively in a timely and coordinated manner, the punishment taken by engine company members in a hostile environment is reduced or eliminated. However, if they fail to do their job as they should, firefighters on the scene will take a beating and the fire loss will be increased beyond what it should be. It has often been said that those chief officers who know how to use their truck companies effectively do the best job on the fireground.

Truck company members are different from engine company members. Training officers attempt to drive all initiative out of rookies during their initial training. The objective is to get them to work and think as a member of a team, as their initial assignments are generally to an engine company where teamwork is essential.

Photo 4.1 Truck companies require apparatus that carry a complete assortment of ladders and sufficient tools and equipment to carry out effectively the responsibilities assigned to the company. (Courtesy of Pierce Manufacturing Inc.)

When assigned to an engine company, a firefighter is normally under the constant supervision of a company officer at an emergency where he or she generally works as a team member in the laying and handling of hose lines. Occasionally he or she works as an individual when manning a small line alone, but this rarely occurs.

The scenario is different when a firefighter is assigned to a truck company. Truck company members are taught to use their initiative and to be able to work and think as individuals. Many times they are on their own while carrying out their duties at an emergency. A truck officer will direct a couple members of his or her company to open up a roof, or to go to the fourth floor and make sure everyone is out of their rooms, or maybe to shut off the utilities. When a truckman is ordered to open the roof, where and what size hole to cut is left to his or her initiative, as is the decision as to how to get the people out or what tools to use to shut off the utilities. Consequently, it is the practice of most departments to assign the more experienced firefighters to the truck companies. Most officers agree that good truckmen are worth their weight in gold.

As a review from Chapter 2, truck companies are assigned the responsibility for:

1. Laddering operations
2. Overhaul
3. Controlling the utilities
4. Ventilation
5. Forcible entry
6. Physical rescue

Photo 4.2 An axe is one of the primary tools of a truckman. (Courtesy of Chicago Fire Department)

The functions for which the truck company is responsible has been presented in the order shown as an aid to remembering the functions. Note that if the functions are reworded slightly, they can be remembered by the word *LOUVER*. (See Figure 4.1.) Although the functions have been listed in the above order to serve as an aid in remembering them, they will be presented more in the order in which they are likely to be performed at a fire.

It is the responsibility of every truckman to be knowledgeable on the how, when, and where to perform each of these functions, and what tools to use to do the job most effectively. Some truck officers assign part of these tasks to individual truckmen as a regular assignment. At a fire, each truckman proceeds to carry out the function assigned him, or her, if not otherwise directed. Assignments should be made so that firefighters work in pairs. Other officers will assign their personnel on a priority system which they establish as a result of their initial fireground size-up.

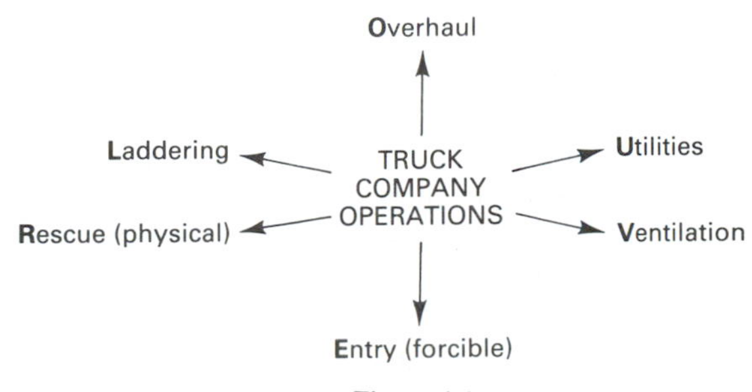

Figure 4.1

PHYSICAL RESCUE

In a limited sense, rescue is the process of removing people from burning buildings, or buildings likely to become involved, to a place of safety. In a broader sense, rescue operations not only include the removal of people but the prevention of further injury and the loss of life at fires. Rescue operations are the number one priority at fires. They are the first consideration that should be given by the first arriving officer at a fire, and should constantly be in the mind of every officer and firefighter on the scene. Some of the questions that should be answered on arrival regarding rescue operations are:

1. Is anyone in the building?
2. If so, are they in immediate danger?
3. If people are in danger, is it possible to rescue them or to prevent them from becoming injured or killed?
4. If so, what is the best method of making the rescue or preventing further injury or loss of life?

Many times the question of whether or not anyone is in the building is answered immediately by cries for help. At other times there is silence, but information can be obtained from persons who have escaped from the building or perhaps from the neighbors. Neighbors are generally a good source of information. They will normally be familiar with how many people live in a house and whether or not the people are home, and will be of great assistance in identifying anyone in the crowd who may have escaped from the burning building. It is good practice to follow up on all information given, but keep in mind that the information is not always reliable.

At other times there are no cries for help and no information available from those at the fire, but the conditions indicate that people are probably in the building. For example, if a unit arrives at a single-family dwelling at 3:00 A.M. and finds the house full of smoke, all doors and windows locked, and no one outside of the building, it is a pretty good indication that people are in the building and that immediate rescue operations are required. Of course, the people who live in the house may be out of town on vacation, but all operations must proceed as if the building is occupied. Every structure should be assumed to be occupied until a complete search proves otherwise.

Occasionally it is apparent when the first unit arrives that several people are in need of rescue. At this time it is necessary for the officer-in-charge to establish mentally a priority list of the order in which people are to be rescued—unless, of course, there are a sufficient number of firefighters at the scene to make all rescues simultaneously. The priority system should be established on the basis of which person is in the most danger. This basically means that an evaluation must be made of who is in the greatest danger and who will probably die first if not rescued. Care must be taken during this evaluation not to let the loudness of the screams nor the begging of relatives, friends, or neighbors influence the decision. It should also be kept in mind that who is likely to die first does not necessarily depend on the order in which the fire will reach the people. Many times panic plays an important part. Due to panic, a person might jump from an upper story while a person who is in much greater danger will remain calm. The emotional state of individual occupants of the building is difficult, if not almost impossible, to determine; however, where it is obvious that panic has placed one person in more danger than another, then this factor must be considered when establishing a priority list.

Care must be taken when establishing a priority list to consider all occupants of the building. It is quite easy to limit an evaluation to those people or conditions that are observed from the front of the building upon arrival. It is not difficult under the stress of an emergency to neglect the fact that a building has four sides. Quite often people in the back of the building who are in more danger than those in front are overlooked because of the problems facing the officer-in-charge at the front of the building. (See Figure 4.2.)

Panic

Panic is one of the primary contributing factors to loss of life when a large number of people are placed in danger at a fire situation. Panic is an emotional reaction to

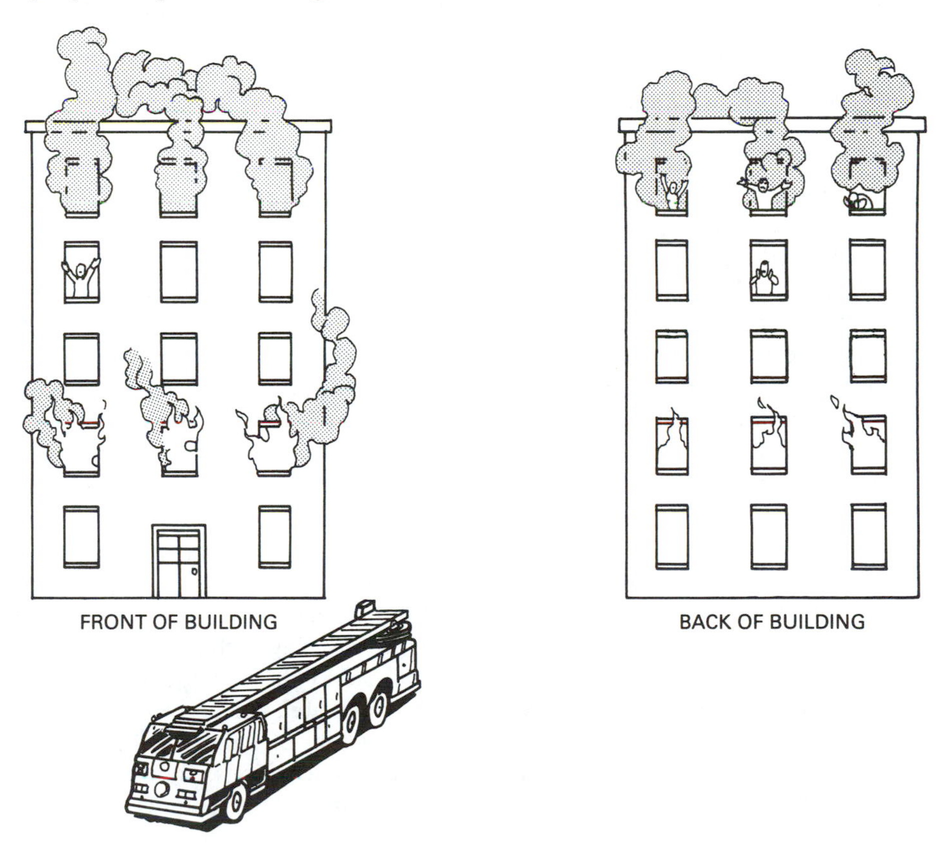

Figure 4.2 Don't forget the back

fear. At times fear robs a person of the ability to reason. A good example is the Cocoanut Grove fire that occurred in Boston in 1942. Nearly 500 people died in this fire. They died because as a group they panicked, and would not stand back and allow room for an inward-swinging door to be opened. Of course, even young school children logically know that a door cannot be opened if several people are pushing their weight against it, yet in a panic situation almost all logical response ceases to exist.

Another factor causing panic at theaters, hotels, and places of public assembly is the occupants' lack of familiarity with the general floor layout. When people are in a strange building, they tend to leave by the way they entered rather than try to determine the quickest and easiest route to the outside.

From a fire rescue viewpoint, it is important to remember that people should be expected to act in an abnormal manner in fear-producing situations. They will jump when they shouldn't jump, they will hide in places they shouldn't hide, they will try to escape from rooms when they should remain there, and they will drag others into danger with them. It does not take a life-threatening situation to trigger panic. Any trifling cause or misapprehension of danger, if allowed to continue, can contribute to panic. It is important early in the emergency to make every effort possible to reassure anyone who appears to be in any degree of danger, and to continue reassuring him or her until a rescue can be made.

Search Procedures

The importance of establishing a priority system for the order in which people should be rescued was previously discussed. Care must be taken not to limit the priority list to those people who can be seen. Many times the people in immediate danger are those who cannot be seen. Most people who die in fires die prior to the arrival of the fire department. However, many die after the department arrives on the scene. At the time of arrival, they may be unconscious, or may not be able to make their presence known due to physical handicaps, or may be trapped. It is important that these people be found quickly. It can best be done through a systematic and thorough search.

The search should commence in the place where people are most likely to be in the most danger. In single-family dwellings during sleeping hours, it's the bedrooms. In multiple-story buildings, it's generally the fire floor, the floor above the fire floor, and the top floor, in that order. (See Figure 4.3.)

One method of ensuring that the search will be conducted in a thorough and systematic manner is to start at the fire perimeter and work toward the center, or make a room-by-room search, thoroughly covering each room before proceeding to the next. Coordination should be established by those conducting the search so that efforts are not wasted by searching the same area twice.

A good search is made by using the senses of sight, touch, and sound. Sound will be particularly beneficial if the victim is conscious. By listening carefully, the searcher may hear weak cries for help, groans, or other sounds originating from the victim. It is good practice to tap a wall or pipes while making the search. This may attract the attention of the victim and thus allow him or her an opportunity to

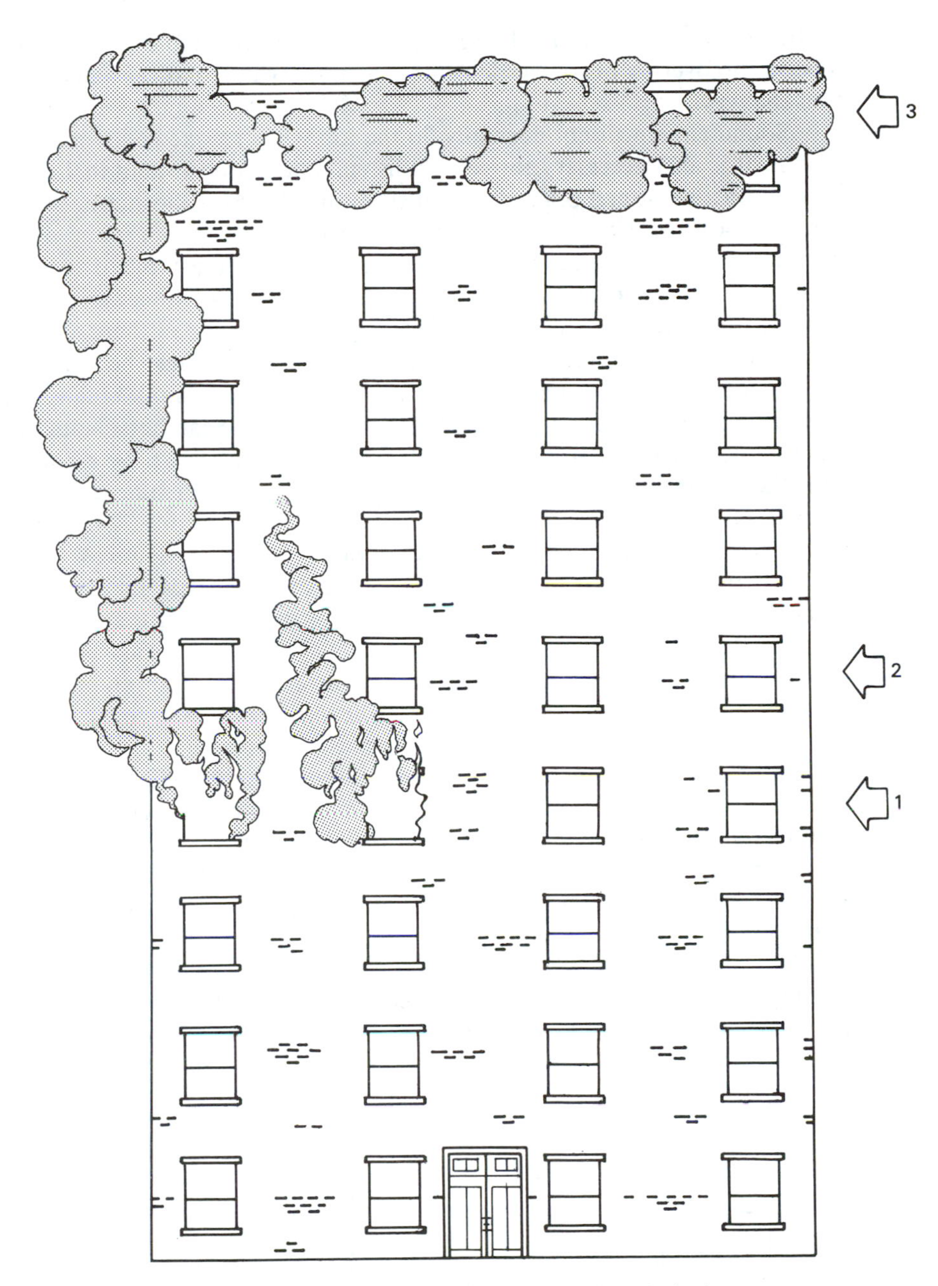

Figure 4.3 Where people are most likely to be in danger

identify his or her location. Even if the individual is unable to make a sound to assist in discovering his or her location, the noise will reassure the victim that help is on the way.

Sound alone will not normally do the job when smoke is down to the floor in a room. It is also necessary to feel and listen. The search should include feeling all chairs, sofas, beds, and other pieces of furniture. Perhaps the victim was occupying one of these when he or she became unconscious. The floor around each piece of furniture should be searched since the person may have toppled over. Searchers should also check around doors and under windows. Victims are often found in these locations because they were able to get only that far in their search for fresh air before they were overcome. The inside of all closets must be searched. People may have entered the closet, thinking it was an exit, or small children may have entered the closet in an attempt to hide from the fire. Searchers should also check under all beds, particularly in children's rooms. Again, they have a habit of hiding from the fire in places they use when playing the game of hide-and-go-seek. The room should be searched thoroughly and quickly, but the firefighter should never leave a room until he or she is positive no one is in it. (See Figure 4.4.)

Every room of a building should be searched. This includes bathrooms, kitchens, basements, and even rooms that have padlocks on them. On more than one occasion firefighters have found children locked in bedrooms. They were placed there by parents who had left the house for the evening. Halls should be given the same consideration as rooms. It is important to remember to check hall closets, as people have been known to enter these thinking they were exits.

In hotels, apartment houses, office buildings, and the like it is a good idea to check the roof. This is particularly important if a stairwell leads to the roof from the interior of the building. People will go to the roof to seek help when escape to the lower floors has been cut off. It is also good practice to observe the roofs of adjoining buildings while making the search. People may have jumped to this nearby area.

The only assurance that the fire department has that all people in danger in a building are out of the building is by way of a thorough and systematic search. It should never be assumed upon arrival that everyone is out of the building, regardless

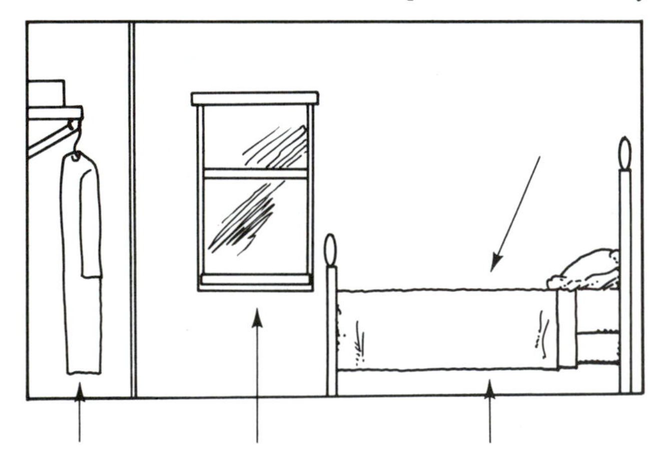

Figure 4.4 Search all areas

of what information is received from people at the fire. This also applies to abandoned buildings. Abandoned buildings are havens for the homeless, and it is good to keep in mind that if a fire starts in an abandoned building, someone was probably inside to start it.

Removal of Victims

When victims are found, they should be carefully removed to a place of safety. How much care is required depends on the condition of a victim when he or she is found. If the victim is seriously injured but not in immediate danger, more time can be taken than if he or she is in jeopardy. If the person has been overcome by smoke and asphyxiation is imminent, it is extremely important to get him or her to a location where resuscitation measures can be started immediately.

The technique used to move a person to a place of safety also depends in part on the victim's condition when he or she is found. If the individual is not injured and able to walk, it may only be necessary to guide him or her to a place of safety. In fact, several people can be taken at one time under these conditions.

In some cases it will be necessary to carry the victim to safety. Generally, one-person carries should be limited to those situations where the victim is not too seriously injured, not too heavy, and the distance to safety is not too great. Where possible, two or more firefighters should be used when it becomes necessary to carry a person to safety. Greater weights can be carried longer distances and it is easier to prevent aggravating an injury. A stretcher or an improvised stretcher should be used if possible, otherwise it will be necessary to use one of the two- or three-person methods of carrying.

Occasionally it will be necessary to remove a victim by dragging. This method is most suitable where the victim is not suffering from any injury and only one firefighter is available for removal. Of course, if the victim is in immediate danger and only one person is available, then every effort should be made to drag him or her regardless of the injuries, the size of the individual, or the distance to safety. It is generally best to drag a person head first. The individual can be dragged by his or her own clothing or by the wrists; however, it is sometimes easier and less hazardous if the person is placed on a turnout coat, blanket, or some other object that can be pulled rather than pulling directly on the victim.

Removal of people from multiple-story buildings presents a much greater challenge than does a removal from single-story buildings. Sometimes it is only a matter of guiding people to safety; at other times their removal may be extremely complex, however. People are more easily moved in situations in which they are familiar. They are used to walking down stairs, but not walking down fire escapes or climbing down ladders. Consequently, when possible, keep them on familiar territory. For example, if people are trapped several stories above the fire floor in a building that has an interior stairwell and a fire escape, but no smoke tower, it is best to guide them down the stairway to the floor above the fire, then out onto the fire escape, and as soon as possible after passing the fire floor, take them back inside and proceed down the stairway.

Stairways are the quickest and easiest means of removing a person from an

Photo 4.3 Although stairways are the quickest and easiest means of removing a person from an upper floor, it is sometimes necessary to use ladders. (Courtesy of Chicago Fire Department)

upper floor. Care should be taken to evaluate the situation completely prior to moving a person toward the ground. For example, if a person is waiting in a window of an upper floor to be rescued when the first unit arrives, it may appear that the best means of effecting the rescue would be to raise an aerial ladder to the window and allow the victim to climb down. However, if it is possible to do so, it would be safer and easier for a firefighter to climb the aerial, go in the window, and guide the person down the stairway.

Elevators should never be relied on as a means of evacuation. It is possible that heat or flame has entered the shaft and weakened the cables or damaged the hoisting machinery. Even if the fire is some distance from the elevator shaft, any interruption in the electrical power could render the elevator useless. In such a case, people in the elevator would not only be placed in danger but would most likely panic.

Using ladders for rescue work should be a last resort. This is particularly true of aerial ladders. Bringing victims down ladders is dangerous. If it becomes absolutely necessary to use ladders for rescue work, it is generally best that the victim climb down on his or her own with a firefighter descending in front of the victim for the purpose of providing guidance and stability.

Summary

Rescue operations take first priority at a fire, but they are challenging and demanding. Many times evacuation and rescue race through the mind of the first officer on the scene when the situation is one where people are in danger. However, it should be kept in mind that in many situations, an immediate and rapid attack on the fire is the best means of saving the lives and preventing further injury of the occupants.

LADDERING OPERATIONS

Laddering operations basically include raising or using ladders for the purpose of making physical rescues, gaining entrance into a building, providing a path for hose lines, gaining access to various portions of the building, and various other phases of firefighting and fire or water control. Both straight ladders and extension ladders are carried on truck companies for performing needed operations. A large range of sizes and types of ladders are carried, varying from smaller ladders, which can effectively be handled by one person, to longer extension ladders, which require the coordinated efforts of six people.

It is not the intent of this chapter to illustrate the various methods of raising the different ladders carried on truck companies. Firefighters should learn this in their initial training as rookies. Rather, it is the intent of this section of the chapter to indicate some basic principles for handling ladders that can be applied to various incidents on the fireground.

Raising Ladders

The best angle for climbing a ladder is 75 degrees. (See Figure 4.5.) This angle makes climbing easier and utilizes the maximum strength of the ladder. Experienced

Photo 4.4 Truck company operations include the raising of ladders. (Courtesy of Wichita, Kansas, Fire Department)

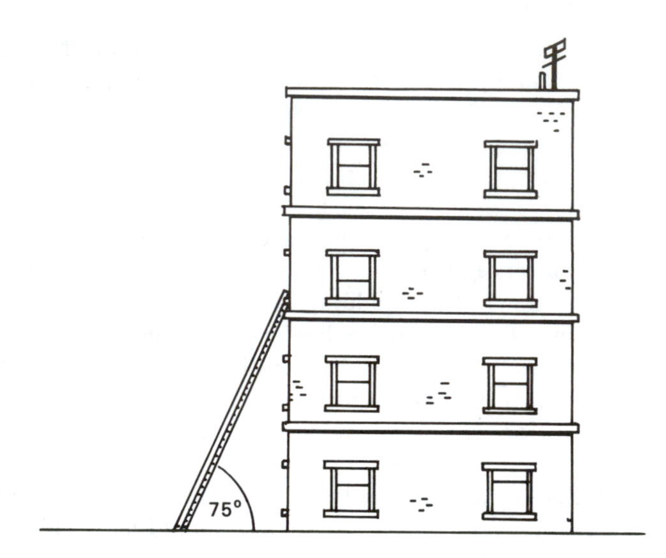

Figure 4.5 The best climbing angle

firefighters do a fairly good job of approximating this angle by making a quick evaluation of the situation. However, there are some guides that have proven to be useful.

The distance that the ladder should be placed from the building in order to obtain a 75-degree angle can be approximated by dividing the length of the ladder by 4. For example, a 20-foot ladder should be placed approximately 5 feet from the building (20 ÷ 4), and a 30-foot ladder should be placed approximately 7½ feet from the building (30 ÷ 4). It should be kept in mind, however, that this method provides only an approximation. Some departments use the system of dividing the length of the ladder by 5 and adding 2. This method provides a slightly different answer, but is just as effective in obtaining fireground results.

When ladders are raised to a window sill for the purpose of making ingress, they should be placed to one side or the other of the window and extend approximately three feet above the window sill (approximately three ladder rungs). Placing the ladder in this position will make it easier and safer for anyone leaving the ladder to enter the building, particularly if the person has a hose line on his or her shoulder. In fact, if lines are taken up a ladder and into the building through a window, it is generally best to place the ladder to the side of the window opposite of the shoulder on which the line is carried. As an example, if it is the practice of a department to carry the hose on the right shoulder, then the ladder should be placed to the left side of the window.

If the ladder is raised to a window for the purpose of making a rescue, it is generally best to place the tip of the ladder at or slightly below the window sill. This position generally makes it easier to remove people from the building.

When ladders are raised to the roof, they should be extended approximately five feet above the roof (about five ladder rungs). This position makes it easier to leave the ladder to gain access to the roof, particularly if carrying a hose line. It also makes it easier to leave the roof, and to see the ladder if departure must be hastened. It is good practice to paint the tips of all ladders white or fluorescent orange. Not only can the white or fluorescent orange tip be seen more clearly through smoke if it becomes necessary to make a rapid retreat from the roof but it also makes it easier to see the end of the ladder when it is being lowered into a window or roof edge. (See Figure 4.6.)

For safety purposes, it is best that all ground ladders be footed when anyone is climbing or descending, unless the ladder is secured to the building. Footing means that a firefighter is stationed at the bottom of the ladder with his or her toes placed against the heel of the ladder or one foot placed on the bottom rung. The firefighter should grasp the beams to help steady the ladder.

It is good practice to secure the ladder to the building at the top if it is to remain in one position for any length of time. This not only relieves a firefighter for other duties but it is also a safety factor in the event someone may find it necessary to climb down the ladder when no one is available for footing.

It should be kept in mind that the 75 degree climbing angle and the distance that the ladder should be above the window sill and roof are ideals; however, these ideals are seldom achieved on a regular basis on the fireground. Generally, the more training and experience the members of a truck company have, the closer these ideals are approached at an emergency. It should also be remembered that safety takes precedence over the ideal angle. For example, better ladder security will generally be provided if the butt of the ladder can be placed in a crack in the cement rather than on a flat, smooth surface. If such a spot is near the intended grounding position, then this should be taken even if there is a slight sacrifice in climbing angle.

Selecting a Ladder

A ladder should be selected that will most closely achieve the ideal; however, there are no set standards as to the location of windows, the distance between floors of

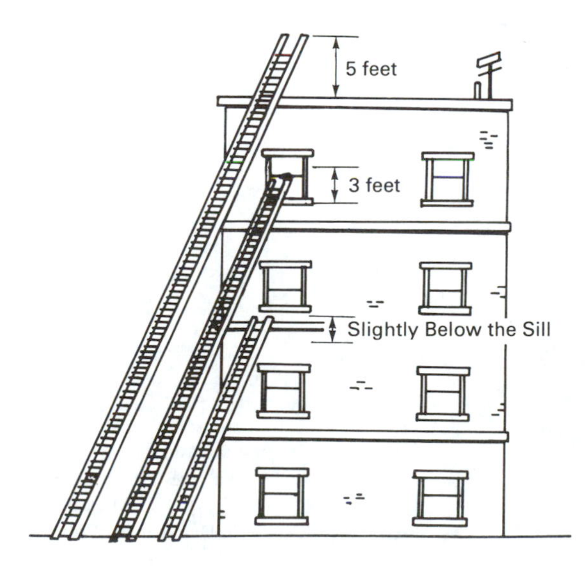

Figure 4.6 The position of the ladder tip will vary with the situation

buildings, or other factors that affect the proper selection of a ladder. Experience, training, and good judgment are required to compensate for the lack of standards.

As mentioned, there is no hard and fast rule regarding the distance between floors of buildings, but there are some guidelines that might be useful in the selection of ladders. The average distance from floor to floor of residential buildings is approximately 9 feet, and the average distance from floor to floor in commercial buildings is 12 feet. Window sills are generally about 3 feet above the floor in both types of occupancies. Parapets on roofs of commercial buildings are generally about 3 feet in height.

One of the reasons that the ideal is seldom achieved on the fireground is that the exact-sized straight ladder required to reach the desired height is not carried on the apparatus. Another reason is that if it is carried, it may not be available at the time needed. Consequently, the best that truckmen can generally do is to come as close as possible to achieving the ideal. The following, although not exact, will provide some guidelines for the selection of ladders.

In a residential building, a 16-foot ladder placed on the window sill of the second floor will extend approximately 3 to 4 feet above the sill. A 24-foot ladder placed on the window sill of the third floor will extend approximately 2 to 3 feet above the sill.

A 20-foot ladder placed on the second floor window sill of a commercial building will extend approximately 4 feet above the sill, while a 30-foot ladder placed on the third floor window sill will extend approximately 2 to 3 feet above the sill.

There is a greater variance in roof heights in buildings than there is distance

Photo 4.5 Ladders are used to advance hose lines into buildings. (Courtesy of Chris Mickal, New Orleans Fire Department Photo Unit)

between floors. This makes it difficult to establish guidelines for ladders to be used to reach the roof. However, more than likely an extension ladder will be used for roof operations. Consequently, some guidelines can be established as to the approximate length of ladder required to reach roofs, and extension ladders can be adjusted accordingly. Keep in mind that the following are only approximations:

	RESIDENTIAL ROOF	COMMERCIAL ROOF
Two stories	24 feet	35 feet
Three stories	35 feet	47 feet

Truck companies carry both straight ladders and extension ladders. Straight ladders are used to best advantage for gaining access to roofs of one-, two-, and sometimes three-story structures, and into second- and third-floor windows. Extension ladders can be used in almost all locations that straight ladders can be used. If there is a choice between using a straight ladder and an extension ladder, the extension ladder should generally be chosen because of the variable height factor. Extension ladders, however, are generally heavier than straight ladders, which may limit their use where manpower is a critical factor.

One other thought should be kept in mind. After selecting and raising a ladder, if it does not reach its intended point, take it down. A ladder left in this position serves no useful purpose and presents a constant threat to firefighters. Someone may try to climb it and fall in the attempt.

Placement of Ladders

It is good practice to ladder all exposures as well as the fire building. Ladders should be raised not only where they are needed at the time but also wherever it is anticipated that they might be needed. However, more ladders than are needed should not be raised at the expense of inside truck work.

Ladders should be raised as close as possible to the point of work. For example, they should be raised to the roof at such locations that firefighters will not have to travel long distances on slippery, steep roofs or over obstacles.

Ladders should not be placed over louvers, windows, or other openings that are likely to emit heat or fire.

Always place ladders in as secure a position as possible. Constantly consider the possible risk to firefighters.

FORCIBLE ENTRY

Forcible entry is the process of using force to gain entrance into a building or secured area. Using force at an emergency becomes necessary when the first arriving unit finds the building secured and entry must be made for the purpose of search and rescue, locating the fire, extinguishing the fire, or carrying out any of the functions required during firefighting or rescue operations. Forcible entry operations may also become necessary once entry is gained if doors or other areas inside the building are found locked or secured.

In most cases, entry into buildings is made through doors or windows, but occasionally it may be necessary to use more drastic means to gain entrance. Also where, when, and how a building is entered will vary from one situation to another. Although there are no set rules that apply to every situation, there are several basic principles that should be considered prior to using forcible entry techniques.

1. Before attempting to gain entrance into a building through a door, regardless of whether it is locked or unlocked, feel the door to make sure it is not hot to the hand. A hot door is an indication that a possible backdraft condition exists inside. Do not open the door under these circumstances until ventilation has commenced.
2. Make sure a door is locked before trying to force it open. This principle may seem basic, but it has been violated a number of times at fires. It is easy to assume that a door is locked when arrival is made at a commercial occupancy at 3:00 A.M. and it appears that no one is in the building.
3. Don't force a door if it is possible to break a small panel of glass and reach inside to open the door. The glass is much easier and cheaper to replace than the door, and this operation generally takes less time than forcing the door. (See Figure 4.7.)
4. Before using force on an entry way, make sure that other possible means of entering have been checked to see if they are open.
5. If time and conditions permit, before using force to gain entry to the first floor of a multiple-story building, check to see if it is possible to raise a ladder to an unlocked second floor window or entry point.

Principles, of course, are nothing but guidelines that must occasionally be violated. If the purpose of gaining entrance is search and rescue, then the primary considerations are time and safety. Entry must be made as quickly as possible, as

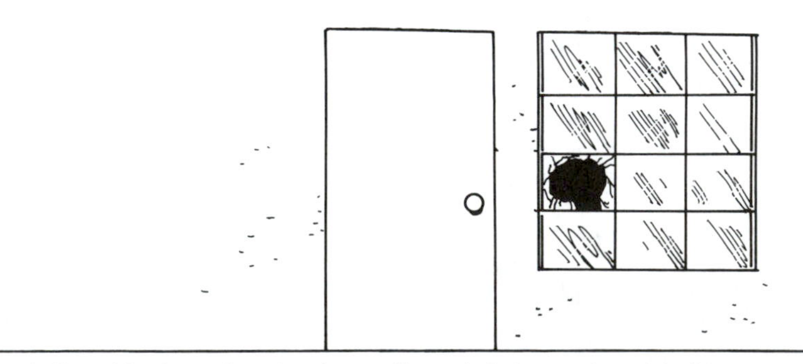

In lieu of forcing a door, break a small panel of glass and reach inside to open the door.

Figure 4.7 Glass is less expensive than a door

even a few seconds may mean the difference between life and death to someone inside.

If entrance is to be made for the purpose of fire extinguishment, then a delay may be indicated. This is particularly true if there are any signs of a potential backdraft. Ideally, entry should be made almost simultaneously with the commencing of ventilation operations, but at actual emergencies this is seldom achieved.

There are many different types of entry ways into buildings. There are also many different types of locks and security devices designed to make unauthorized entrance into buildings more difficult. Sometimes entrance is a simple matter of breaking a window, twisting off a padlock with a claw tool, or using an axe to force a single-hinged door. At other times, getting inside is slow and complicated. The extra care taken by occupants to secure their building against intruders has brought about the need for special tools and special knowledge to gain entrance into some buildings. Fire companies carry electric power tools, hydraulic tools, and gasoline-powered tools to assist in forcible entry operations. It is important that those firefighters responsible for forcible entry become familiar with the different problems that exist within their district, and learn to use the tools necessary to accomplish the objective.

CONTROLLING THE UTILITIES

Control of the utilities should take place early during an emergency. This is important not only for the potential saving of property but as a safety factor to the firefighters working at the scene.

Utility shutoffs may be located almost anywhere in a building. The ideal time to locate the various shutoff locations is, of course, during a pre-fire planning inspection. Not only should the location of the shutoffs be found and identified but the area of the occupancy that individual shutoffs control should be understood. Unfortunately, most fires occur in buildings that have not been subjected to a pre-fire planning inspection. In these cases it is generally necessary to cut off the utilities where they enter the building rather than taking the risks involved with delaying the shutoffs until interior control points can be found.

Water

Broken water pipes within the building can cause unnecessary loss and could possibly cause a restricted water supply for firefighting purposes. It is best to shut off the water system to the building if there are signs of a significant leak or broken pipes in the system.

Water systems in buildings generally have a shutoff where the supply enters the building. In residences, this shutoff can usually be found at the outside faucet nearest the street. If it cannot be found quickly, then it is usually best to shut off the water at the street. There is usually a cover in the sidewalk in front of the building that will identify this location. Generally, but not always, the water is on when

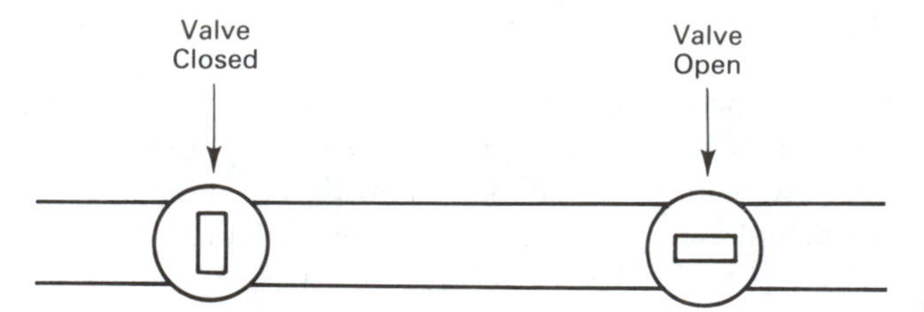

Figure 4.8 Position of valve shutoffs

the valve lines up with the pipe and off when the valve is perpendicular to the pipe. (See Figure 4.8.)

Gas

The fuel supply to the building should be shut off if there is any possibility that there are broken pipes or a leak inside the building. Sometimes the broken fuel pipe or leak will be feeding the fire, at other times it will be setting up an explosive potential within the building. Gas entering the building can be shut off at the meter. There is no standard location for gas meters. In single-family residents the gas meter is generally outside. Whether the gas is off or on can usually be determined by the position of the valve. (See Figure 4.8.)

Some properties use LPG for heating and cooking. Storage tanks are almost always outside and are easy to identify. Shut off of the supply to the building can be made at the storage location.

Electricity

Burned or broken electrical wires are a potential life hazard to firefighters. If there is any possibility that such a condition exists within the building, then electricity to that portion of the building should be shut off. If the location of an inside control panel is known, and positive information is available as to the portion of the building controlled by the various circuit breakers, then it is safe to shut off the electricity at that location. However, if there is any doubt as to the location of the control panel or what portion of the building it controls, then the electricity should be shut off as it enters the building. To effect a positive shut off at this point, it is generally best to cut the wires as they enter the building. The cut should be made at the supply loop as it enters the building. (See Figure 4.9.). However, *all cutting of wires should be done only by a reliable person from the electrical department.*

In the event that there is no loop as the power enters the building, it may be necessary to have the electrical department cut the wires from the pole to the building. In this case, the wires should be cut as near to the pole as possible. Cutting the wires close to the building would result in "hot" wires falling to the ground where they would create a severe hazard. Prior to having the wires cut, it is important to make sure that no personnel are in the area where the wires will fall. Do not let anyone come in contact with the first wire that is cut until the second has been cut, for there is a possibility that the first wire could be "hot." If three wires enter the

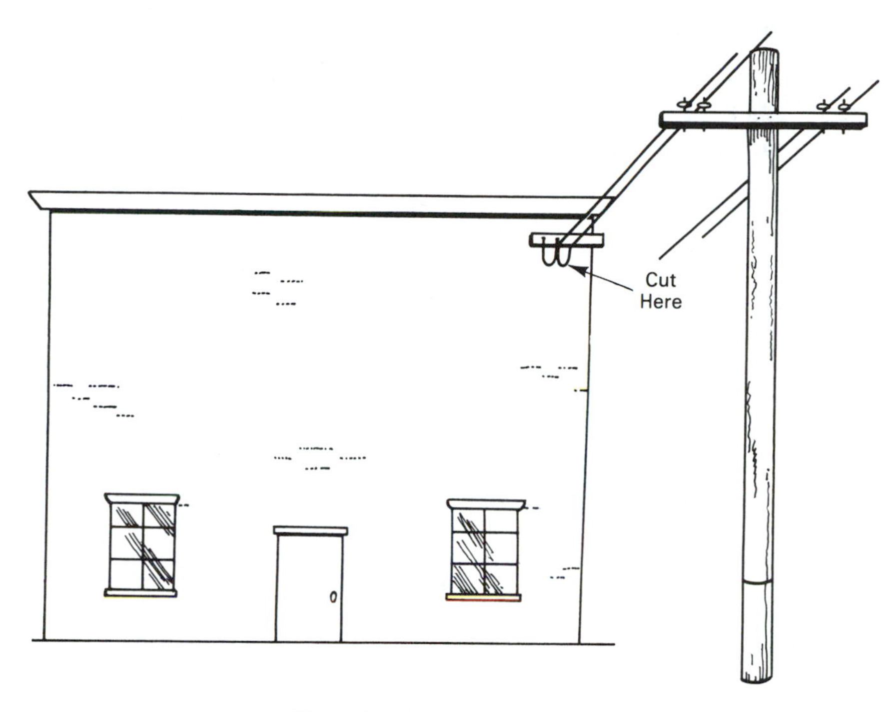

Figure 4.9 Cut wires at loop

building, all must be cut and the first two should be considered "hot" until the third has been cut.

VENTILATION

Ventilation is a process of replacing a bad atmosphere with a good atmosphere. As applied to the fire service, it means clearing a structure, vessel, or other area of hostile smoke, heat, or noxious gases through controlled channels and replacing the objectionable gases with fresh air.

Proper ventilation has been called the key to successful firefighting. Stated in reverse, failure to ventilate in the proper place and at the proper time, or improper ventilation, has been termed one of the greatest and most common failures of firefighting operations. Some of the favorable results of proper ventilation are:

1. Firefighting and rescue operations are facilitated.
2. The danger of backdraft is reduced.
3. The area is made more comfortable for firefighting personnel.
4. The survival profile of people in the building is increased.
5. The spread of the fire is curtailed.
6. Toxic and explosive gases are removed from the area.

7. "Mushrooming" is eliminated.
8. The fire will burn more freely and the volume of smoke will be reduced.
9. Heat and smoke damage will be held to a minimum.
10. Fire, smoke, and gases can be properly channeled.

To be most successful, ventilation operations should be given careful consideration during the size-up of the fire. The work should start immediately once the decision is made as to where ventilation should take place. Although good firefighting techniques dictate that ventilation should not be started until sufficient hose lines are in place both to confine and extinguish the fire, from a practical standpoint, it is almost impossible to ventilate before hose lines have been laid. Roof ventilation, for example, requires that ladders be raised, firefighters ascend to the roof, the proper location determined as to where ventilation should take place, and a hole cut in the roof. These steps take much more time than does the laying of lines. Additionally, engine companies generally arrive on the scene prior to the truck companies and have a head start on operations.

There is no standard as to the type and amount of ventilation required at a fire. Each fire is different, and the conditions at each fire will dictate what is needed. Where possible, vertical ventilation is preferred to horizontal ventilation since smoke normally rises and seeks vertical outlets. This does not mean, however, that a hole should be cut in the roof at every fire. The height and construction of the building, along with the volume of smoke and the location of the fire, will dictate the type of ventilation required.

Photo 4.6 Truckmen preparing to ventilate the roof of a two-story dwelling. (Courtesy of Wichita, Kansas, Fire Department)

Roof Ventilation

Roof ventilation will vary from a simple task to an almost impossible one. Operations might be one of simply removing a scuttle hole cover to a complex one of manually opening a large area. One of the first thoughts when preparing to ventilate a roof is to look for natural openings that can be used. Of course, these openings are of little value unless they are in the right location. If they are not in the correct location and are opened, they will pull fire to the opening. Some of the natural openings available for ventilation are scuttle holes, skylights, and stairwells. (See Figure 4.10.)

A scuttle hole is an opening in the roof that extends into the attic. It is fitted with a cover that is sometimes locked. Scuttle holes can normally be used to ventilate only the attic, whereas some of the other natural openings can be used for additional ventilation. When lifting off scuttle hole covers, it is good practice to have the wind to your back so that the heat and gases will be released in a safe direction. A wooden cover can generally be removed by prying it open with an axe. Spring-loaded metal covers that are locked from the inside present more of a problem and sometimes it is better to look for another location for ventilation rather than try to open them.

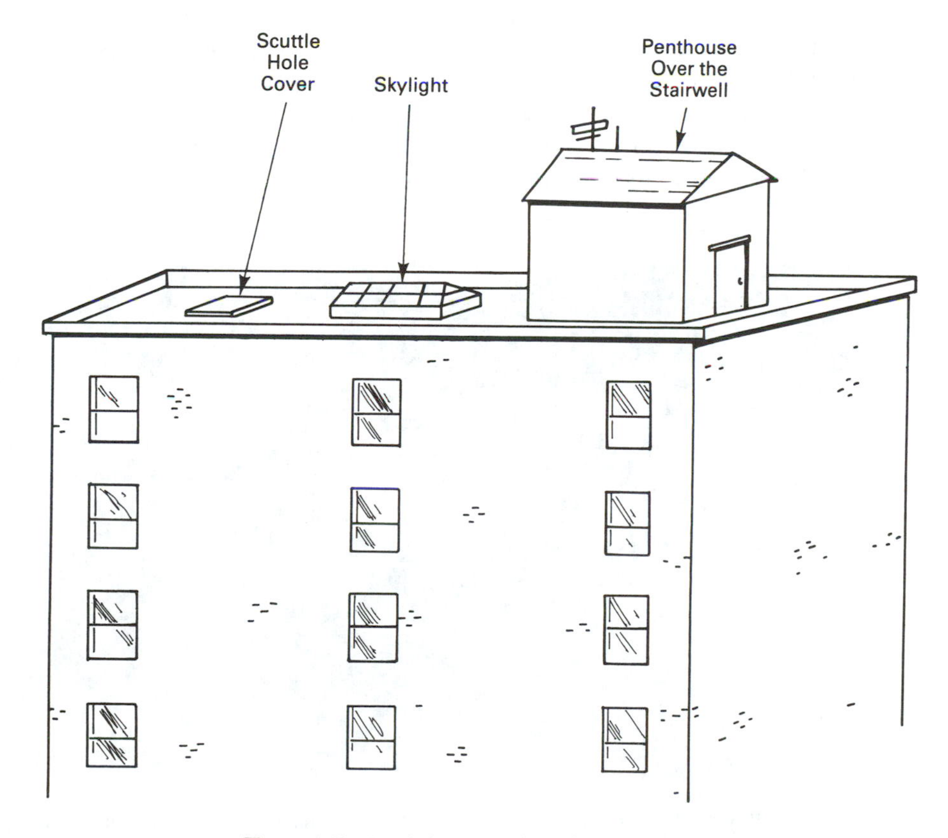

Figure 4.10 Look for natural ventilation openings

Skylights are used to provide light to areas below the roof. Their value for ventilation, as is so with other natural openings, depends on what area they will ventilate. In hotels and apartment houses they will generally ventilate the hallway of the top floor. In commercial buildings they will generally ventilate the manufacturing area.

Skylights can generally be removed intact since the frames are normally only lightly secured at the corners. However, if any difficulty is encountered in removing them, and time is important, there should be no hesitation in breaking the glass. Some skylights are equipped with plexi-glass rather than wire glass. The plexi-glass will melt at approximately 400° F., which normally provides self-ventilation.

The use of a stairway that terminates on the roof for ventilation can be particularly valuable; however, it can also prove to be dangerous if improperly used. It is a simple operation, as it only requires opening a door. Previous knowledge of the building at this point can prove to be invaluable. The information gained can be used to ascertain what part of the building below the roof will be affected by opening the door. It should be remembered that extensive ventilation operations, particularly where they affect the life hazard, allow little room for error.

Two factors are of paramount importance when it becomes necessary to cut a hole in the roof. The first is that it be properly located. This means that the hole should be cut directly over the fire, *providing it is safe to do so*. Sometimes this spot is readily identified by simple observation. If not, look for discoloration of the roof or feel the roof for the "hot spot." (See Figure 4.11.)

Cutting a hole in the ideal location means that the place where the hole will be cut is usually the hottest and most dangerous place on the roof. If there is any doubt as to the safety of personnel, then this spot should be avoided and a safer

Photo 4.7 Mechanical equipment can be used effectively for ventilation. (Courtesy of Chicago Fire Department)

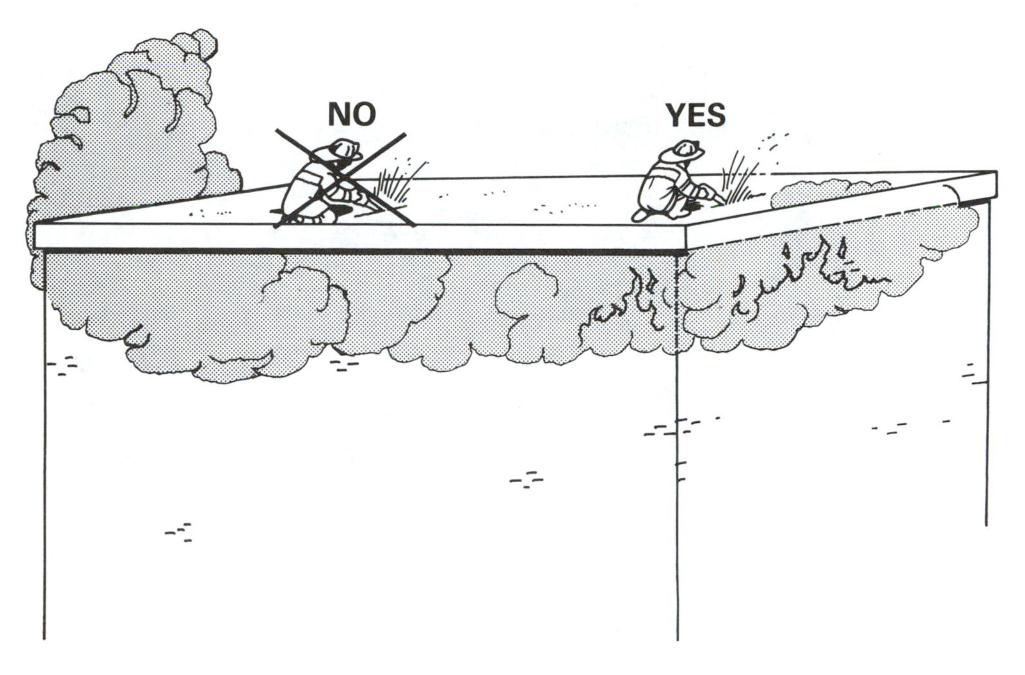

Figure 4.11 Be careful where the hole is cut

location selected to make the cut. It is normally possible in a large building to cut the hole within 20 feet of the hottest spot without seriously pulling the fire through previously uninvolved portions of the building. Pulling the fire through uninvolved portions of the building should normally be avoided; however, there should be no hesitation to do so to protect life. There also should be no hesitation to do so when working on roofs of lightweight construction or other roofs that are a hazard to firefighters.

An evaluation of the roof should be made to ensure that it is safe before any firefighter is allowed on it. As firefighters enter the roof they should sound it out ahead of their movement to locate any weak areas. A rubbish hook or other long-handled tool can be used for the sounding. A good knowledge of roof construction is important in making this evaluation.

Some roofs, by their very nature, are hazardous and require extreme caution when working on them. An example is a roof covered with ½-inch plywood on a commercial building, which gives the appearance of being sturdy. Most roofs on buildings constructed prior to 1960 are solidly constructed and well supported. They are generally safe, although they might give the impression of being somewhat "springy." Extra care, however, should be taken when working on buildings constructed after 1960 that are classified as lightweight construction. Commercial buildings so classified generally have flat roofs with ½-inch plywood covering. Fire will rapidly weaken large sections of these roofs, exposing firefighters working on the roof to extreme danger. If fire is impinging on the roof, operations should be conducted close to the fire but over an uninvolved portion of the roof. It is not safe to

Photo 4.8 Fire from this apartment has extended into the attic. (Picture taken prior to roof ventilation.) (Courtesy of Wichita, Kansas, Fire Department)

Photo 4.9 Fire after attic has been ventilated. (Courtesy of Wichita, Kansas, Fire Department)

operate over any portion of the roof that is involved with fire. If the roof is obviously too "soft" or "spongy" to be safe, then all personnel should be removed from the roof. Generally, under these conditions, the fire is about to "self-ventilate." When it does so, no attempt should be made to extinguish it as it comes through the roof, for it is doing exactly what would have been done with proper manual ventilation.

Working on a roof, directly over the fire, is a dangerous operation. Many firefighters have gone through the roof and been killed during ventilation operations. When conditions are right, elevated platforms can be used as a safety factor. A line can be lowered from the platform and tied to the person opening the roof. The firefighter can be pulled to safety in the event the roof gives way. (See Figure 4.12.)

A roof with truss construction is an excellent example of where this method of safety may be particularly important. These roofs are inherently dangerous. It is a good idea to watch for the use of "gangnail" plates during any pre-fire planning inspection.

The second factor to consider when cutting a hole in the roof is that the hole should be of sufficient size to vent the fire adequately. There is no set rule as to the size of hole to be cut. The intensity and size of the fire, the rapidity of its spread,

Figure 4.12 Think safety

the necessity for immediate ventilation, and the type of building construction all play an important part in making a decision. It is good practice, however, to adopt the principle of overventilating rather than following the natural tendency to underventilate. A 10′ by 10′ opening on the roof of a large building with a well-involved fire inside would not be considered out of line. Remember, a 10′ by 10′ opening will provide the same ventilation area as four 5′ by 5′ openings. Not only would the larger opening be more effective but it would be easier and cheaper to repair than four smaller openings.

Louvering is an effective method of ventilating a roof. (See Figure 4.13.) It is a system that helps reduce the exposure of personnel to smoke and heat as the roof is vented. It can be used effectively on sheathing but is most effective on plywood paneled decking.

The operation consists of making two longitudinal cuts on either side of a single rafter with the cuts parallel to one another. The longitudinal cuts are then intersected with cross cuts. Once the cuts are made, one side of the cut panel can be pushed down, resulting in a louvering effect. This operation will ventilate the area below and direct most of the heat and smoke away from the firefighters conducting the operation. Cuts should be made so that the firefighters making the cuts have the wind to their back as the panel is louvered.

Make sure the roof is able to be penetrated before starting to cut a hole in it. Axe blades have been dulled by firefighters attempting to cut through concrete roofs.

Some roofs are covered with tar paper or tar and gravel coatings. These mate-

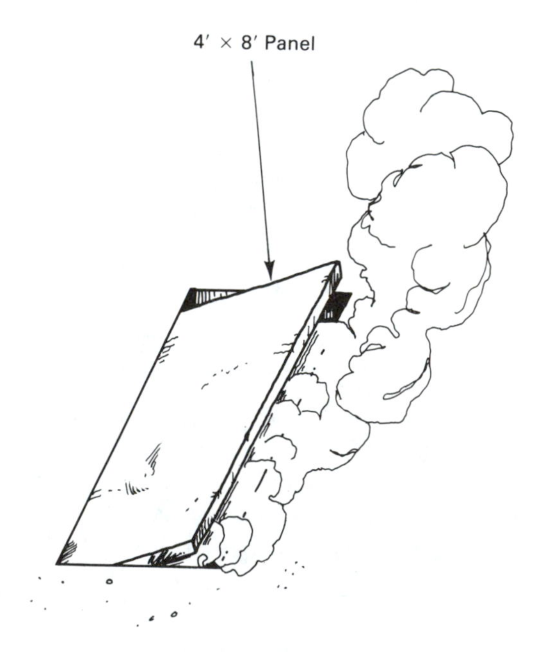

Figure 4.13 Louvering is an effective ventilation method

rials should be removed before cutting a hole. The pick-head of the axe will be effective for this operation.

When holes are cut in roofs, cuts should be made along the rafters to facilitate making repairs. (See Figure 4.14.) The rafters can usually be located by sounding with the axe. Rafters provide a dull sound as compared with the higher-pitch sound of the spacing between rafters. Cuts can be made with an axe or a chain saw. The holes should be oblong or square. All cutting should be made before any sheathing is removed. The firefighter removing the sheathing should stand to the windward side of the hole and start removing the sheathing from the leeward side, working toward the windward side. This will protect the crew from the heat and smoke. If the hole is of the proper size and the job of ventilating has been done correctly, the rafters will not be damaged as the sheathing is removed.

The windward side is the side from which the wind is blowing, and the leeward side is the side toward which the wind is blowing. For example, if the wind is blowing from the north, then the north side of the building would be the windward side and the south side of the building would be the leeward side.

If plywood sheathing is used on the roof, the operation can be simplified by removing an entire panel intact. When large areas are involved, at least two adjacent panels should be removed and three would not be excessive.

If the hole has been properly located, and there is nothing below to block its path, heat, smoke, and fire should start issuing from the opening as the sheathing is removed. It is always best to have a charged line on the roof at this time to protect exposures from the resultant heat. If there is a minimal amount of smoke, heat,

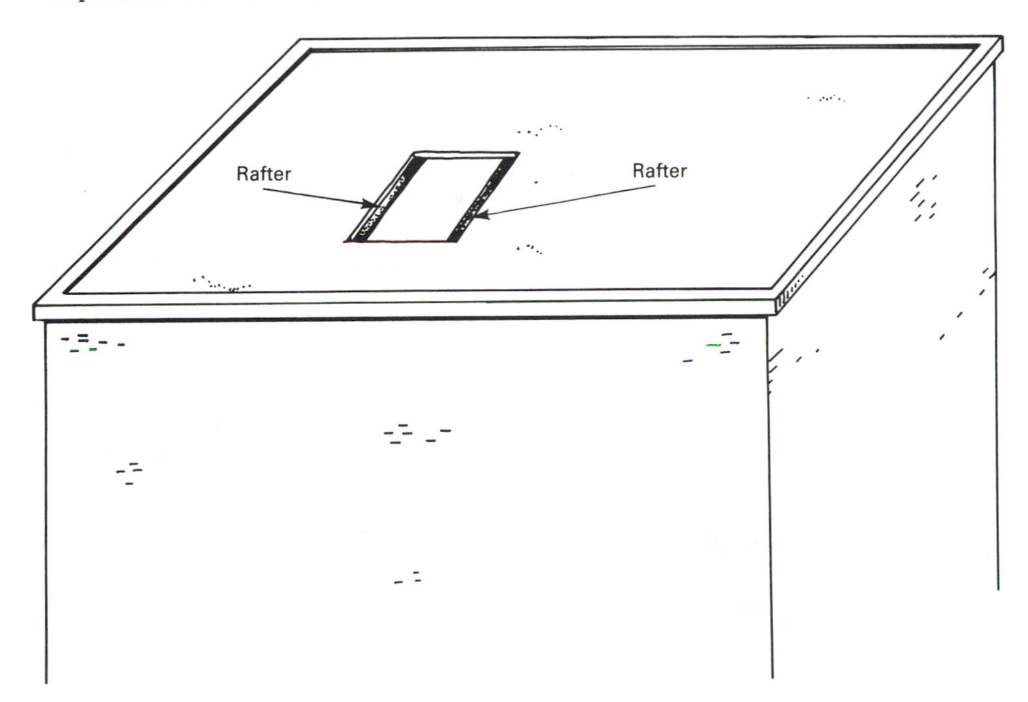

Figure 4.14 Make cuts along rafters

and fire issuing from the opening, it is probably due to restrictions to flow below the roof. If the fire is not in the attic, the restriction will probably be a ceiling. This can usually be corrected by using a pike pole or similar object through the hole to open the ceiling below.

One last thought on roof ventilation: Timing is important. Although the ideal is seldom achieved, it should be strived for and the results of early or late opening visualized. The ideal is to open the roof a few seconds prior to the engine company making entrance at the lower level and entering the area with charged lines. The opening of the roof will start a movement of the built-up heat, gases, and smoke from the interior of the building to the outside through the opening made. When the opening is made for the engine company to enter, fresh air will be pulled into the room, which will accelerate the outward movement of the heat, gases, and smoke. This will generally clear the interior atmosphere sufficiently to allow hosemen to make a rapid advance on the fire. (See Figure 4.15.)

If the roof is opened early, movement of the built-up heat, gases, and smoke will commence due to the natural tendency of heat to rise and the pressure buildup inside the building; however, it will be somewhat restricted. The movement will be accelerated once an opening is provided at a lower level. Early opening provides more safety for hosemen advancing lines than does late opening.

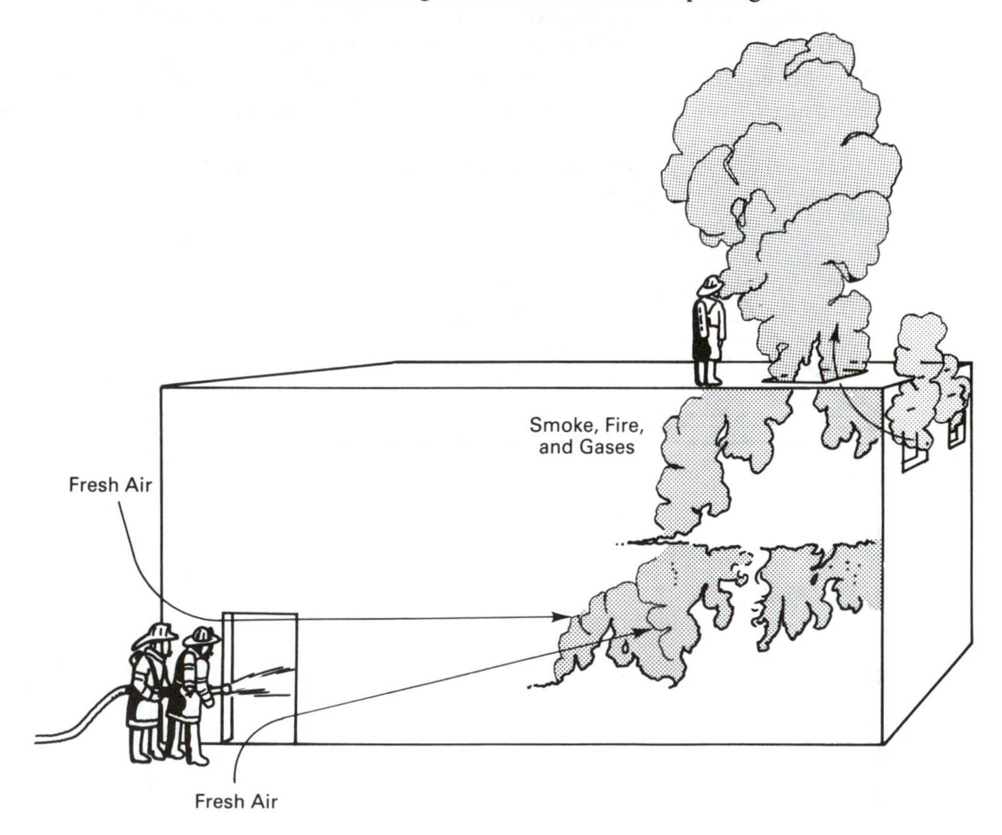

Figure 4.15 Proper timing helps the hosemen

Late opening is the least desirable of the three but probably occurs most often due to the time required for truckmen to ascend to the roof and make the opening. Under these circumstances, hosemen are required to advance lines into an atmosphere with a heavy concentration of heat and smoke. However, their working conditions are immediately improved once the opening is made.

Use of Vertical Openings

Since the natural tendency of fire and heated gases is to rise, built-in vertical openings provide an excellent means of venting the fire. However, care must be taken when selecting which vertical openings to use, as a bad choice could increase the hazard to life or spread the fire. The three most common vertical openings that can be used for ventilation are stairwells, elevator shafts, and light wells.

Before using a stairwell for ventilation, make sure that all occupants of the building are below the fire floor and be certain that the stairwell will not be used for egress. The stairwell can be used most effectively if it extends through the roof. Operation only requires that the door to the penthouse over the stairwell be opened at the roof; however, some action may be required to ensure that the door will remain open.

Additional steps must be taken if the stairwell terminates at the top floor. It will be necessary to cut a hole in the roof over the stairwell or to cross-ventilate the top floor. It may be necessary also to cross-ventilate the floor below the top floor if the smoke is extremely heavy. Smoke ejectors blowing up the stairwell from intervening floors will help facilitate the removal of smoke. (See Figure 4.16.)

Enclosed elevator shafts can also be used effectively for vertical ventilation. Before attempting to use the shaft for ventilation, make sure the cage is below the fire floor. This will then provide a clear opening from the fire floor to the upper terminating point of the elevator. However, if the cage is secured at some point above the fire floor and cannot be moved, and it is imperative or extremely beneficial to use the elevator shaft for ventilation, then additional steps must be taken. The elevator doors at the floor below the secured cage must be opened and this floor cross-ventilated. Again, if the smoke is extremely heavy, cross-ventilation should be repeated at the next floor below. (See Figure 4.17.)

Most elevator shafts terminate at the top floor. This means that the top floor and possibly the floor below the top floor will have to be cross-ventilated if the shaft is used. Whatever has to be done at the upper levels should take place prior to the heat and gases being allowed to enter the elevator shaft.

Light wells are vertical openings found in older multiple-story buildings that are used to provide light and air to enter rooms of the building. Some times the light well extends from the ground floor to the roof, and at other times it will commence somewhere on an upper floor and extend to the roof. Make sure that all windows opening into the light well are closed before using these vertical openings for ventilation. Although these openings can prove to be very effective for removing smoke from a building, they can be dangerous if an attempt is made to pull fire up the opening. The fire could break the windows at intermediate floors, which would result in a rapid, uncontrolled extension of the fire.

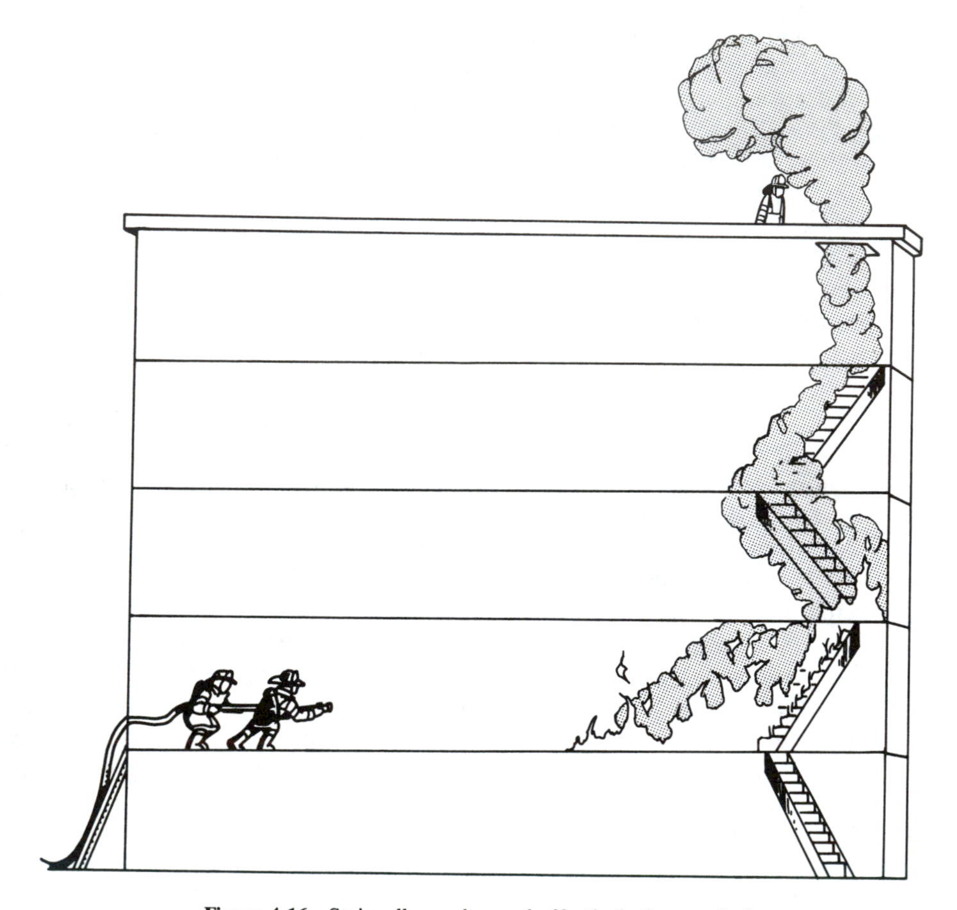

Figure 4.16 Stairwells can be used effectively for ventilation

Some light wells are covered at the top of the opening with a skylight. The skylight should be completely removed for maximum effectiveness.

Cross-Ventilation

Cross-ventilation is a means of using windows, doors, and other horizontal openings for ventilation. This method can be extremely effective whenever there is a large number of openings of sufficient size. It has the advantage of normally limiting the smoke to the floor being ventilated, which is particularly beneficial when a life hazard is present or valuable merchandise in other areas of the building are susceptible to smoke damage.

An evaluation should be made of the wind direction prior to commencing cross-ventilation operations whenever an entire floor or a large area is to be ventilated. The operation will be most successful when the smoke and gases are channeled out the leeward side of the building. Once the direction of the wind is determined, double-hung windows on the leeward side should be opened from the top

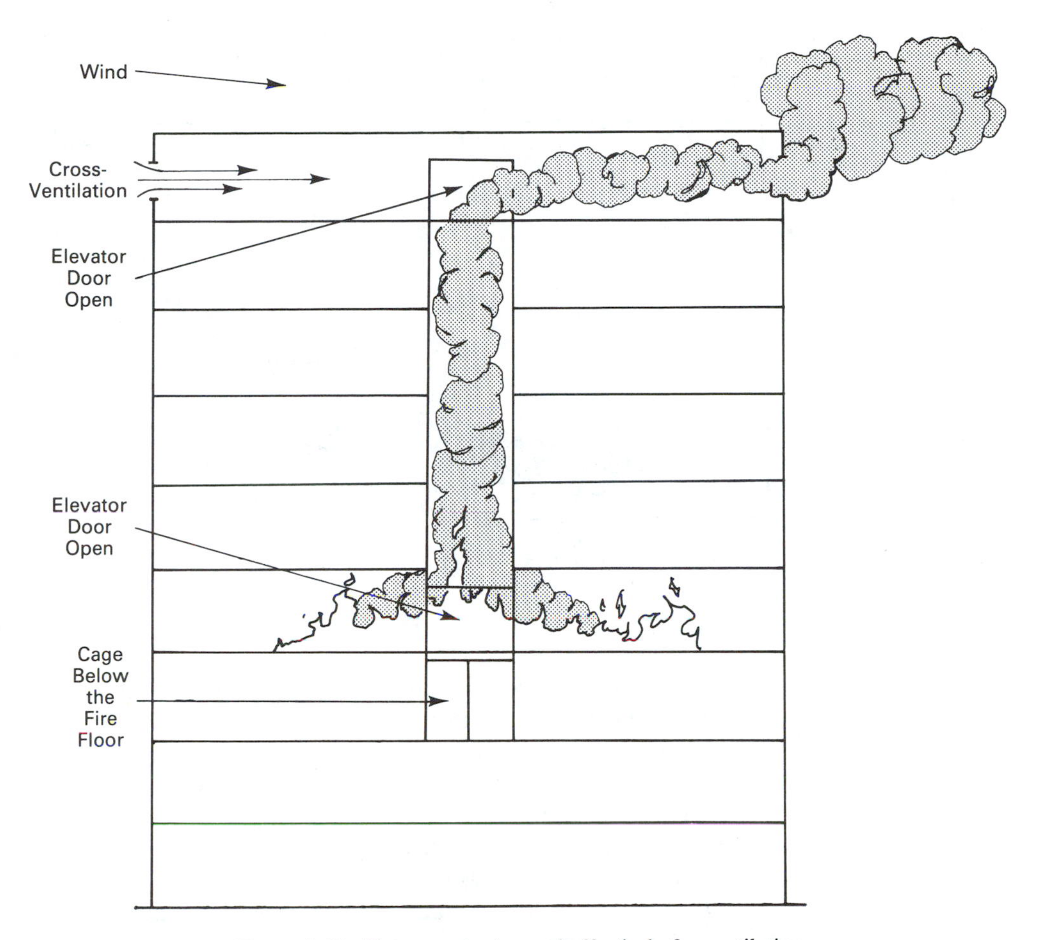

Figure 4.17 Elevators can be used effectively for ventilation

and sliding aluminum or sliding wooden windows should be fully opened. Remove all drapes, shades, screens, and other obstructions that would impede the flow. Double-hung windows on the windward side should be opened from both the top and the bottom or from the top only after all the leeward windows have been opened. Sliding aluminum or sliding wooden windows should be fully opened. Again, all obstructions to flow should be removed. (See Figure 4.18.)

Occasionally the air will be still and no horizontal movement can be established. In this case, double-hung windows should be opened two-thirds from the top and one-third from the bottom. Sliding aluminum or sliding wooden windows should be opened to provide for maximum air movement. This method should also be used in small rooms where horizontal openings are restricted to one or two sides of the building.

It is good practice when cross-ventilating to determine where the heated gases and fire will go once they leave the openings. It is possible that they could extend upward and through open windows on the floors above or through openings into adjoining buildings. If there is such a possibility, then all exposed unprotected open-

Figure 4.18 Effective use of cross-ventilation

ings should be closed or spray streams used at these openings to prevent any extension of the fire.

Sometimes it is desirable to cross-ventilate where it is impossible to get to the windows that need to be opened. In this situation it may be necessary to break the windows. This can be accomplished from the inside by the use of hose streams. Hose streams may also be used outside for breaking windows, or they may be broken by the use of pike poles, ladders, or a ball and chain. Pike poles can sometimes be used from ground level or it may be necessary to operate from the floor above the fire, using either pike poles or the ball and chain. If ladders are used, the ladder should be lowered into the window in such a manner that the glass will fall inside the building. Do not extend the ladder into the window. If this is done, panels of glass may slide down the ladder and seriously injure those on the ground. Once the window is broken, the ladder should be removed as fire is likely to shoot through the opening.

Smoke ejectors can be extremely useful as an aid in cross-ventilation. They can be set up to create either a suction or blower effect. When used to create a suction effect, they will pull the smoke toward the blower and out the opening. When used as a blower, they will move the smoke and gases toward openings. They can also be used in pairs, with one providing a suction action and the other acting as a blower.

Positive Pressure Ventilation

Positive pressure ventilation is a method of clearing an area of smoke and gases for the purpose of gaining entry to extinguish the fire. It combines the processes of making openings and using mechanical blowers. For maximum effectiveness, it requires good timing. The theory of the positive pressure ventilation method can best be demonstrated by referring to Figure 4.19.

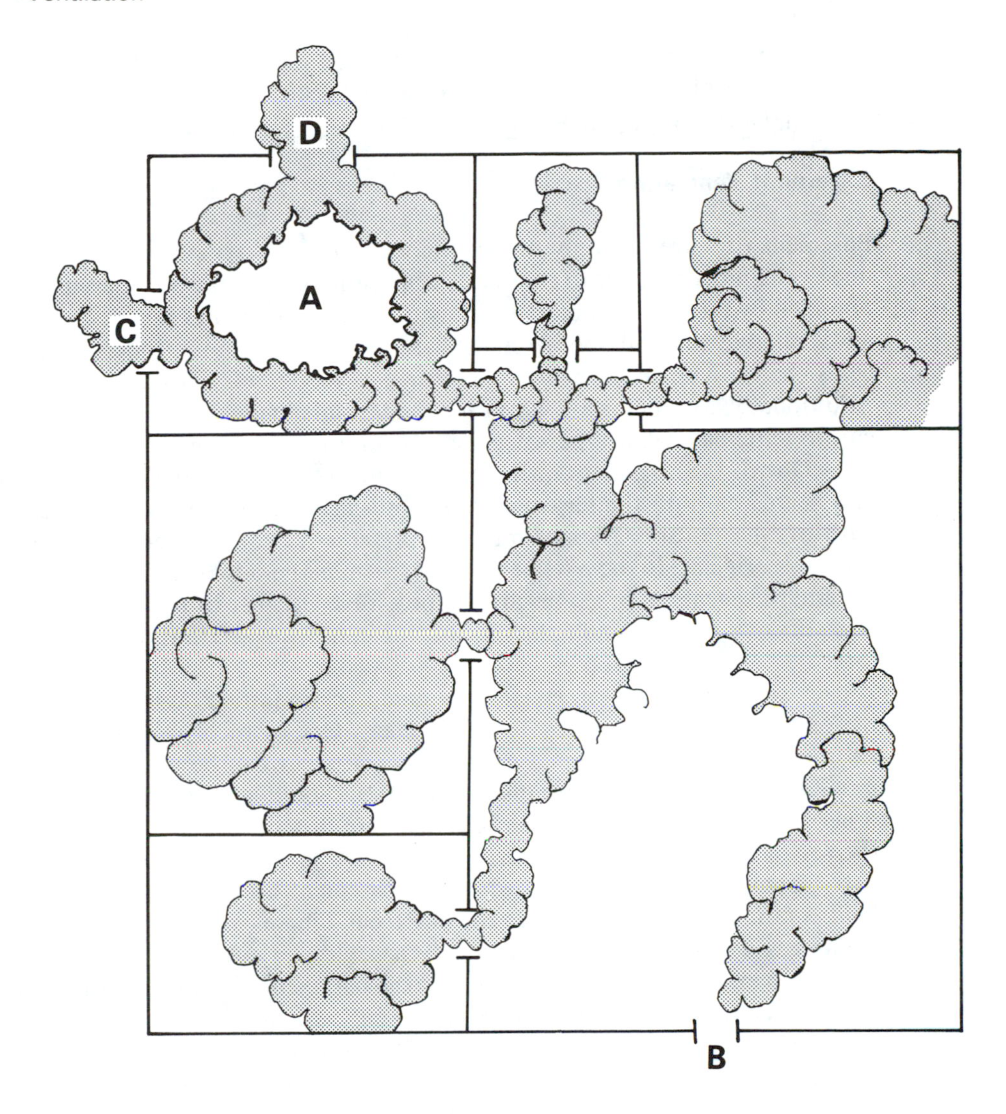

Figure 4.19 Forced ventilation

The fire is confined to room A and heavy black smoke has penetrated all areas of the house. Firefighters are positioned at the front door (B), ready to enter the building. As the truckmen break out the windows at points C and D, two firefighters with blowers enter the front door ahead of the hosemen, forcing fresh air toward the fire area. The combination of the openings at points C and D and the forced air quickly clear the area of smoke, permitting the firefighters to enter for extinguishment. The blowers should continue to be operated after the fire is extinguished to aid in the removal of residual steam and smoke from the fire area. Following are two points that should be considered regarding positive pressure ventilation:

1. Do not attempt to use the positive pressure ventilation method for entry if the interior of the building is ripe for a backdraft.
2. Make the exhaust openings in the fire area as close to the fire as possible.

Basement Ventilation

Basement fires are not only one of the most difficult types to extinguish but also one of the most difficult to ventilate. The ventilation problem of basements is compounded by the fact that most basements have few openings directly to the outside and a relative small number from the basement to the ground floor. In fact, in some cases the only opening to the basement is a single entry from the ground floor. Additionally, basements are used for storage, which means that it is extremely difficult to anticipate what types of materials will be involved in the fire.

One of the objectives in fighting basement fires is to confine the fire, smoke, and heated gases to the basement area until means can be provided to control the path of their escape. Many times this presents a problem because some vertical openings originate in the basement, which permits the smoke and fire to rise through the walls if adequate fire stops are not provided. Because of the amount and different types of storage found in basements, it can be expected that the fire will produce large quantities of smoke.

Ventilation of the basement area might be simplified if there are openings to the outside at opposite ends of the basement. One of the openings can be used to rid the basement of unwanted gases while the other can be used to bring in fresh air. The opening used for exhaust of the gases should be opened first. Entering the other opening with blowers will help facilitate the movement of the gases, as will the use of spray or other streams.

Sometimes light is provided to the basement area through the use of dead lights in the sidewalk adjacent to the basement. These can be broken to provide an opening, but this is a slow and difficult task.

If there is but a single opening into the basement, or only one of the openings available is suitable, it may be necessary to cut or breach a hole in the ground floor to provide ventilation. (See Figure 4.20.) This hole should be made next to a window or door opening so that the vented gases can be directed to the outside of the building. Blowers or spray streams can be used effectively to assist the movement of the gases through the opened window or door. When using spray or fog nozzles to assist in ventilation, operate the nozzle at approximately a 60 degree angle and direct it in such a manner that it will cover about 85 percent of the opening. (See Figure 4.21.) An adequate number of charged lines should be laid out prior to opening the hole to protect against the spread of fire once ventilation commences.

Positive pressure ventilation can be used effectively to remove the smoke from the basement. One or two fans can be used at the stairway entrance into the basement.

One last thought on ventilation of basements: Due to the confined area and the tremendous amount of heat that can build up, basement fires are often ripe for a backdraft. It is good practice to consider that a backdraft will occur if adequate precautions are not taken to prevent it.

Figure 4.20 One method for basement ventilation

Figure 4.21 Spray streams can be used effectively as an aid in ventilation

Additional Ventilation Practices and Principles

Following are some general practices and principles that should be considered during ventilation operations:

1. Breathing apparatus should always be worn during ventilation operations. The types of materials used in building construction, as well as in storage, may produce toxic gases that can prove dangerous even when an atmosphere appears relatively clear.
2. It should be expected that the fire will increase in intensity once an opening is provided to the outside. Charge lines should always be in position to protect exposures and horizontal fire spread across the roof prior to providing the opening.
3. Firefighters should always work in pairs when ventilating a roof of considerable size or where large volumes of smoke are present. In fact, when sufficient personnel are available, it is good practice to have firefighters work in pairs in all phases of ventilation operations, with a company officer present if possible.
4. A plan of escape should always be established prior to making an opening. At least two ladders should be raised to the roof if roof ventilation is to take place. This will provide a second means of escape in the event the planned route is cut off.
5. Unless life is in danger, or there is no alternative, do not provide openings that will pull the fire unnecessarily through uninvolved portions of the building.
6. Do not provide openings that will endanger adjacent property unless adequate lines are provided to protect these areas.
7. Do not ventilate a roof after the need has passed. Roof ventilation should serve a useful purpose, and not be done automatically or as an afterthought.
8. Always expect and be prepared for a backdraft.
9. Always consider the direction and velocity of the wind when contemplating ventilation operations. The wind can greatly aid or hinder operations, depending on how plans are made for its use.
10. Delay ventilation of special types of fires, such as ship fires or fires involving electrical equipment, until overall fire strategy has been established.
11. Do not direct hose streams into openings provided for ventilation. Such operation not only eliminates the purpose of providing the opening but will subject crews working in the interior of the building to unnecessary punishment.
12. The operator should be positioned "out-of-line" of the saw when using a chain saw for cutting a hole in a roof. The cut should be made through the sheathing only. Never cut into trusses or other structural supports.
13. Use good common sense when ventilating. Try to provide openings that will keep damage to a minimum. However, do not let the fear of damage be the determining factor as to where and how to ventilate. This is particularly true when life is at stake. Damage should be of little concern in such cases.

OVERHAUL

Overhaul is the final task performed by firefighters at the scene of a fire. Although the primary objective of overhaul is to ensure that the fire is out, it generally includes doing whatever is necessary to leave the premises in as safe and secure a state as possible.

The amount and degree of overhaul work done at a fire will vary considerably from one department to another, and many times will vary from one district to another within the same department. Some departments and some chief officers consider that their job is to extinguish the fire and leave. Others look at overhaul operations as an excellent opportunity to build good public relations. Concepts projected in this portion of the chapter are based on the second principle.

Objectives

There are four primary reasons for considering good overhaul practices as an essential part of firefighting operations.

1. The primary purpose is to ensure that the fire is out. If the fire is not completely extinguished it might result in a rekindle. A rekindle is a situation where the fire department leaves the scene, thinking they have extinguished the fire, and then are called a second time to the location because they had not completely extinguished the fire the first time. Sometimes the fire is larger when they return than it was upon their initial arrival. A rekindle is an unpardonable sin. It leaves a stigma on the record of any officer and is an embarrassment to all firefighters who worked on the fire. Consequently, it is important that any remaining fire be located and completely extinguished before the department leaves the scene.

2. Eliminate any additional water damage that could be caused by extinguishing operations or the weather.

3. Make sure that all portions of the building are accessible to anyone entering from the outside. This means that passageways must be provided so that the owner, insurance adjuster, or any other person having a legal right to enter can inspect all portions of the premises.

4. Make sure that all portions of the building are left in as safe a condition as possible. Operations to accomplish this objective include removing hanging timbers, piping, and wiring; pulling down unsafe walls or chimneys; and clearing all windows and doors of broken glass.

Planning Overhaul Operations

Overhaul operations should normally be started as soon as the fire has been brought under control and adequate personnel are available. However, it may be necessary to delay these operations until an investigation has been made by the arson squad if the fire is of a suspicious nature. Once operations have commenced, they should not be carried out in a haphazard manner. Planned operations should be so well thought out that they can be accomplished in as systematic and professional manner as possible.

The first step in the planning process is for the officer-in-charge to size-up the situation to determine what has to be done and the best way for getting it done. The size-up generally begins by the officer-in-charge making a "walk through" of the premises to make sure that it is structurally safe for work parties to enter and commence working. During the walk through the officer-in-charge should determine where it would be best to store salvageable material and which spot would be best for dumping debris. Some of the other factors that should be determined during the survey are:

1. Are the firefighters at the scene physically capable of doing the work? The stress and physical exertion required during the extinguishment phase may have fatigued personnel to the point where it would be best to call for fresh crews to do the overhaul work.
2. What is the condition of the exterior walls and chimneys? Do they present a hazard to people who may approach or walk next to the building or adjoining property? If so, the condition must be corrected.
3. Are there any unsafe conditions that should be corrected? Consideration should be given to broken glass in windows or doors, loose-hanging timbers or ceilings, wet plaster on the ceiling, holes in the floor, or unprotected vertical openings such as burned-away staircases.
4. Is there a water removal problem? If so, what will be the best way to eliminate it?
5. Is there equipment on the premises such as forklifts or skip loaders? Are there employees on the premises who are trained in the use of the equipment that might be able to assist?
6. What tools and equipment will be needed to complete the work? Are they available on the scene or will they have to be ordered?
7. Does the department have special on-call equipment that might be useful during this particular overhaul operation? Thought should be given to the possible need for an air wagon to refill air bottles, if one is not already on the scene. If the department does not have the needed special equipment, it may be available from the established mutual aid plan.

Once the survey has been completed, or perhaps prior to making the survey if time permits, the officer-in-charge should attempt to contact the owner of the premises if he or she is not at the fire. This is particularly important if the fire is in a commercial occupancy. It is always best that the owner be on the premises before overhaul operations commence in the office area.

Putting the Plan into Operation

When a plan of operation has been formulated and the officer-in-charge is assured that there is adequate personnel and the necessary tools and equipment to do the job, the plan can be put into operation. It is normally better to divide the available personnel into work groups and place each work group under the command of a responsible officer. Sometimes the necessary work groups are already divided by company, under the command of an experienced company officer. The work groups should be assigned to the various tasks that must be done or to various portions of the building in such a manner that the work can be done freely, without everyone getting in each other's way.

It is good practice to continue the use of blowers during the overhaul operations to keep the carbon monoxide level below the acceptable level. The working atmosphere should be checked with carbon monoxide monitoring instruments to confirm that the carbon monoxide level is below the acceptable limit prior to the removal of breathing apparatus. If the officer-in-charge is unable to confirm that the atmosphere is below the acceptable level, then all personnel should wear breathing apparatus during the overhaul phase.

Care must be taken during the overhaul operations to make sure that unnecessary additional damage is not done. Salvageable material must be separated from the unsalvageable and placed in a predetermined location.

Particular care must be taken when conducting overhaul operations in an office area. If the owner has been contacted but has not arrived on the scene, overhaul operations should be limited to ensuring that the fire is out. Care must be taken to save any partially burned records, as they may contain the information needed to determine the inventory, the amount of insurance carried, and the financial condition of the business. The information contained in the records may be vital to the owner in proving his or her loss, and may be invaluable to arson investigations in the event the fire is suspicious.

During cleanup operations the firefighters should be on the alert for any valuables that might be mixed in with the debris. This is particularly important in residential fires. The importance of setting aside partially burned items should not be overlooked. Many insurance companies make a habit of not paying off on insured personal property unless there is proof that it was destroyed during the fire. The meager remnants of a burned object, such as a fur coat or a painting, may be all the evidence an owner needs to collect for his or her loss.

The amount of work necessary to ensure that the fire is out will depend a great deal on the burning characteristics of the material involved. Debris should be separated on the basis of burning characteristics. Material such as wood can be collected and thoroughly wet down as it is removed from the building and thrown into a pile. Smoldering mattresses and overstuffed furniture should be removed from the building and overhauled outside. Fires in these objects are deep-seated. Once they have been taken outside, any burned material should be removed and immersed in a bucket or tub of water. Such material should not be piled with other debris until it is certain that no fire remains. Caution should be taken in removing material capable of deep-seated fires from upper floors of buildings. If an elevator is used for this purpose, it is best to wrap the object in such a manner as to eliminate air getting to the fire area. It is also best to have a water-type extinguisher in the elevator. On several occasions firefighters have been threatened by a flare-up of a

Photo 4.10 Firefighter checking for extension of the fire during overhaul operations. (Courtesy of Wichita, Kansas, Fire Department)

deep-seated fire as an inrush of fresh oxygen fed the fire when the elevator descended.

If space is available to conduct the final extinguishment portion of the overhaul process inside the building in large industrial or mercantile occupancies, it may be advantageous to do so. This does not, however, eliminate the need for the burned material to be turned over and thoroughly wetted down. In residential occupancies it is generally best to move the debris to the outside of the building. Regardless of the choice, the spot where the debris will be piled should be chosen prior to any movement. It is unwise to have to handle the burned material twice. If it is decided to pile the debris outside the building, a spot should be chosen that is a safe distance from the building and on the fire building property unless it is impossible to do so. The proper authorities should be notified if it becomes necessary to pile the debris on the sidewalk or street so that adequate barricades may be set up. Small lines should be laid out and at least one firefighter left to man it at the location where the debris will be piled. The debris should be turned over and thoroughly wet down as it is removed from the building and thrown on the pile.

Many times removing the debris becomes a problem. With some fires, it may take longer to remove the debris and wet it down than it did to extinguish the fire. Most truck companies carry rubbish carriers that can be used for removing the debris. This operation is cumbersome and normally ties up four people. If the overhaul operation is extensive, it is best to utilize wheelbarrows and even skiploaders if they are available. Regardless of what is used, it is better that the debris be removed from the premises in a safe manner rather than being thrown out.

It should be remembered that removal of debris from a fire is for the purpose of making sure the fire is out. If there is an extensive amount of debris inside a building, and the officer-in-charge is sure the fire is out and the premises are safe, then the removal of the debris is the responsibility of the owner. Under these circumstances, removal of the debris should be under the instruction and supervision of the building owner. The owner may use his or her employees, or contract with a private company that is properly equipped and trained in this type of operation.

Overhaul operations include the removal of water from a building, but the removal of all water in the building is not the responsibility of the fire department. The objective of water removal from the building is to prevent further damage. For example, water puddles on a concrete floor will not normally cause any additional damage, whereas the same size puddle on a hardwood floor could cause extensive damage if not removed immediately. Discretion should be used by those responsible for overhaul operations, as arguing who is responsible for the removal of a small amount of water does not necessarily improve public relations. Water removal is more thoroughly discussed in Chapter 5.

There are a number of additional thoughts that should be considered during overhaul operations.

1. All avenues through which heat may have extended during the fire should be thoroughly checked for hidden or smoldering fires. Attic scuttle holes should be located and a firefighter sent into the attic to check for hot spots. Do not overlook the tight space next to the eaves. Birds have a habit of building nests there. The nests could have become ignited without having come into direct contact with any flame.

2. Check all concealed spaces, particularly walls and under floors between the joists. Fires in the walls can be located by running a hand over the wall. A hidden fire is indicated if a spot is found that is too hot for the hand. If such a spot is found, the wall should be opened and the fire extinguished. Continue to remove the plaster, wall board, or other covering until all char to the wood is visible. After wetting down the charred area, it is generally best to use a small amount of water from a spray nozzle and direct it upward, allowing the water to cascade down the inside of the remaining wall.

If fire is suspected between floors, it is generally best to go below the suspicious spot and pull the ceiling. Ceilings are easier and cheaper to repair than floors. Of course, if fire is suspected under the floor and it is impossible to get below that floor, then do not hesitate to cut a hole in the floor.

Fire has a habit of getting under facings around windows, around doors, and at floor level. If fire is suspected in these locations, baseboards and the facings around the doors and windows should be removed. Whenever upper floors have been involved with fire, the bottoms of all vertical openings, including elevators, should be checked for burning debris.

3. Broken glass left in windows, doors, transoms, and so on presents a hazard to anyone entering the premises. Consequently, effective overhaul operations demands that all broken glass be removed from these areas. All broken glass on floors, sidewalks, streets, and the like should be swept up and safely discarded.

4. If not done during firefighting operations, the electricity should be shut off to the burned-out area if the insulation on any electrical wiring has been burned off. If this cannot be done by pulling a switch or circuit breaker at the main control station, then it may be necessary to have the wires to the building cut. If this is done, make sure that proper authorities are notified. Gas and/or water lines should also be shut off if there is any evidence of leakage from these sources.

5. Be on the alert during the overhaul process for any material that is subject to spontaneous heating upon becoming wet. If this material has become wet during the fire, it should be taken outside and laid out in thin layers to prevent the build up of heat.

6. Take particular care when overhauling bales of cotton. Fire has a way of boring deep into a bale, making extinguishment difficult. Water will not normally penetrate to the seat of the fire due to its high surface tension. Wet water can be used to assist in extinguishment but it will reduce the salvage value of the cotton. The best way of making sure the fire is out is to tear open the bale. Cut the bale wires at the center first, then the outside wires. This will permit better control of the contents by limiting the amount of cotton strewn about.

7. Be cautious when overhauling material having a nitro-cellulose base. These materials are capable of producing a flash fire when heated, due to the fact that they contain oxygen in their chemical makeup. It is best to immerse the material in a container of water.

8. Restrict the opening of packaged material to those packages that have been damaged by water or fire. Packaged material might be readily salvageable, which will help hold the loss to a minimum.

9. Fires in multiple-story buildings may make it necessary to throw material out the window to the ground below. Before starting such operations make sure that the space below is clear and that neither people nor property will be placed in jeopardy by such action. However, remember that it is good practice not to throw anything out the window that can be carried out.

10. Early in the overhaul process, any heavy objects such as baled cotton or similar material should be moved from the center of the room toward the wall or placed over supporting columns. Floors may have been weakened by the fire and the water absorbed by the material during the firefighting phase will have increased the weight. This combination could lead to floor collapse if corrective action is not taken. If it becomes necessary to remove some of these heavy objectives from the building, a window frame can be removed and the wall taken out to the floor level.

Securing the Building

If possible, the building should be turned over to the owner when the department is ready to leave. If the owner is not present and cannot be contacted, it then becomes necessary to secure the building against unwanted visitors. All windows and doors that were broken during firefighting operations should be boarded up. Lumber can usually be found on the premises for this purpose. If not, interior doors can be removed and nailed over openings.

The building should also be made as safe as possible from weather elements before leaving. If holes were cut in the roof or if the fire burned through the roof, the holes should be covered with tar paper or plastic if there is a possibility of rain, snow, or other weather elements that might cause further damage.

At some fires it may be necessary to establish a fire watch. This can be done by leaving a firefighter on the scene with a portable radio. If this is impractical, then the first-in company can make a "drive by" a couple of times an hour to check for possible rekindles. It may be wise for the company to stop and reenter the building to check for any smoke developing.

REVIEW QUESTIONS

1. What is the difference between a truck company and a ladder company?
2. What is the primary difference between a firefighter assigned to a truck company and one assigned to the engine company?
3. What are the six functions for which a truck company is responsible?
4. What word can be used to remember all six of these functions?
5. In a limited sense, what is rescue?
6. In a broader sense, what do rescue operations include?
7. What four questions should be answered regarding rescue operations when arriving on the scene of a fire?
8. Who is generally considered as a good source of information as to whether or not people are in a fire building?
9. What is necessary when arriving at a fire and finding several people in need of rescue?
10. On what basis should a rescue priority system be established?
11. What is panic?
12. From a fire rescue situation, how should it be expected that people will act in fear-producing situations?
13. Do most people who die at fires expire prior to or after the arrival of the fire department?
14. When planning to make a rescue search, where should the search commence?
15. In what location are people most likely to be in danger in single-family dwellings during sleeping hours?
16. On what floor are people most likely to be in danger in multiple-story buildings during a fire?
17. What method should be used to ensure that a search will be conducted in a thorough and systematic manner?
18. What senses should be used to make a good search?
19. How is sound beneficial when making a search?
20. What parts of a room should be searched when conducting rescue operations for overcome people?

21. What is the only assurance that the fire department has that all people in danger in a building are out of the building?
22. Why is it necessary to search so-called abandoned buildings?
23. What is it that normally determines how much care should be taken in removing a victim to a place of safety?
24. What should be the limiting factor in using the "one-person carry" to remove a victim from a building?
25. When is the method of dragging most suitable for removing a victim from a building?
26. What method should be used to drag a person from a building?
27. What is the quickest and easiest means of removing a person from an upper floor?
28. What part should elevators play as a means of evacuation?
29. What part should ladders take for use in rescue work?
30. What is the best way to use an aerial ladder for rescue work?
31. What should be remembered regarding using an attack on the fire in association with rescue work?
32. What does ladder operations basically include at a fire?
33. What is the best angle for climbing a ladder?
34. How does one determine the distance from the building that the ladder should be placed to obtain the best climbing angle?
35. Where should the ladder be placed when it is raised to a window sill for the purpose of making ingress?
36. How far above the window sill should a ladder be placed when it is raised for the purpose of making ingress?
37. Generally where is the best location to place a ladder when it is raised to a window for the purpose of making a rescue?
38. Approximately how far above the roof should ladders be extended?
39. What is the advantage of painting the tip of a ladder white or fluorescent orange?
40. What is the average distance between floors in residential buildings?
41. What is the average distance between floors in commercial buildings?
42. Generally how high are the parapets on roofs of commercial buildings?
43. Approximately how far above the sill will a 16-foot ladder extend if placed on the window sill of the second floor in a residential building?
44. Approximately how far above the sill will a 24-foot ladder extend if placed on the window sill of the third floor in a residential building?
45. Approximately how far above the sill will a 20-foot ladder extend if placed on the window sill of the second floor of a commercial building?
46. Approximately how far above the sill will a 30-foot ladder extend if placed on the window sill of the third floor of a commercial building?
47. Approximately what length ladder is required to reach the roof of a two-story residential building?
48. Approximately what length ladder is required to reach the roof of a two-story commercial building?
49. Approximately what length ladder is required to reach the roof of a three-story residential building?

50. Approximately what length ladder is required to reach the roof of a three-story commercial building?
51. Where can straight ladders be used to best advantage?
52. What is meant by forcible entry?
53. When should force be used to gain entrance into a building?
54. What should be done before attempting to gain entrance into a building through a door?
55. What should always be done before using force to open a door?
56. What should be done before using force to gain entry to the first floor of a multiple-story building?
57. Where can the shutoff for the water supplied to a building generally be found?
58. When should the fuel supply to a building be shut off?
59. Where can the gas entering a building generally be shut off?
60. When should the electricity to a portion of a building be shut off?
61. When and by whom should the wires that supply electricity to a building be cut?
62. Where and by whom should the wires to a building generally be cut?
63. How and by whom should the wires to a building be cut if there is no loop where the wires enter the building?
64. What is ventilation?
65. What does ventilation mean as applied to the fire service?
66. What are some of the favorable results of proper ventilation?
67. When should ventilation operations start?
68. What standard has been established as to the type and amount of ventilation required at a fire?
69. What should be one of the first things to consider when preparing to ventilate a roof?
70. What are some of the natural openings available for roof ventilation?
71. What area is normally ventilated by scuttle holes?
72. How can wooden scuttle hole covers normally be removed?
73. What area will skylights generally ventilate in hotels and apartment houses?
74. What area will skylights generally ventilate in commercial buildings?
75. How can skylights generally be removed?
76. What two factors are of paramount importance when it becomes necessary to cut a hole in a roof?
77. How can the place on the roof that is directly over the fire be determined?
78. Where is the ideal place to cut a hole in the roof?
79. What will probably be the result if a hole is cut in the roof at any location other than the ideal location?
80. What is the construction of a roof classified as lightweight construction?
81. What should be remembered about working on roofs of lightweight construction?
82. How can an elevated platform be used as a safety factor when cutting a hole in a roof?
83. What is the natural tendency regarding the size of hole to cut when ventilating a roof?
84. What is meant by louvering as a means of roof ventilation?
85. How can the rafters in a roof generally be located?
86. Which is the windward side of a building?

87. When ventilating a roof with plywood paneling, how many panels, as a minimum, should be removed when large areas are involved?
88. What is the ideal timing for opening a roof?
89. Which is generally better—a late opening of the roof or an early opening of the roof?
90. What are the three most common vertical openings that can be used for ventilation?
91. What precaution should be taken before using a stairwell for ventilation?
92. What must be done if a stairwell is to be used for ventilation and the stairwell terminates at the top floor?
93. What precaution should be taken before using an enclosed elevator shaft for ventilation?
94. What should be done if the elevator cage is secured at some point above the fire floor and cannot be moved and it is imperative or extremely beneficial to use the elevator shaft for ventilation?
95. Where do most elevator shafts terminate?
96. What are light wells?
97. What should be done before light wells are used for ventilation?
98. To what side of a building should the heat and smoke be channeled when using cross-ventilation?
99. How should the windows on the windward and leeward side of the building be opened, and which should be opened first, when cross-ventilating?
100. How should the windows be opened when cross-ventilating and no horizontal movement of air can be established?
101. During firefighting operations, how can windows be broken out from the inside for ventilation if it becomes necessary?
102. How should ladders be used for breaking glass for ventilation?
103. What is meant by forced ventilation?
104. What are a couple of points that should be considered regarding forced ventilation?
105. What is it that causes the primary problems in basement ventilation?
106. How should a basement be ventilated if there are openings at opposite ends of the basement?
107. What might be necessary in order to ventilate a basement if there is but one opening into the basement?
108. At what degree angle should the nozzle be set and how much of the opening should be covered with the stream if spray or fog nozzles are used to assist in ventilation?
109. Why should breathing apparatus always be worn during ventilation operations?
110. When should firefighters work in pairs during ventilation operations?
111. What should be remembered regarding ladders as related to roof ventilation?
112. When should backdrafts be expected when preparing to ventilate?
113. How should the operator be positioned when using a chain saw for cutting a hole in a roof?
114. What is the primary objective of overhaul operations?
115. What are the four primary reasons for considering good overhaul practices as an essential part of firefighting operations?
116. What is a rekindle?

117. When should overhaul operations commence?

118. What is the first step in the planning process for overhaul operations?

119. What are some of the factors that the officer-in-charge should determine during the overhaul survey?

120. How should work groups be divided during the overhaul process?

121. How should debris be separated during the overhaul process?

122. What precautions are necessary when overhauling smoldering mattresses and overstuffed furniture?

123. What should be done if it is necessary to take material that is subject to deep-seated fires from upper floors to the outside by way of elevators?

124. What should be done if it becomes necessary to pile debris on the sidewalk or street?

125. If possible, what should be done regarding removal of debris if the overhaul operations are extensive?

126. When does the removal of the debris become the responsibility of the owner?

127. What is the objective of the fire department removing water from a building?

128. What areas in particular should be checked for hidden or smoldering fires?

129. How can fires in walls be detected?

130. What should be done if fire is suspected between floors?

131. What type of care should be taken when overhauling bales of cotton?

132. How should material having a nitro-cellulose base be overhauled?

133. What practice should be established regarding throwing things out the window?

134. What should be done with heavy objects such as baled cotton during the overhaul process?

135. What should be done if it is not possible to turn the building over to the owner before leaving?

CHAPTER 5

Salvage Operations

OBJECTIVE

The objective of this chapter is to examine the various phases of salvage operations. Review questions are provided at the end of the chapter to assist the reader in determining his or her level of comprehension of the chapter contents.

The firefighting procedure of many fire departments prior to 1900 was to go in the front door with heavy lines and drive the fire out the back door. Many fire officials gave little thought to water damage. In their opinion, their job was to extinguish the fire and go home. If it took a lot of water—so be it.

Insurance companies, on the other hand, were quite concerned about the reckless use of water by the fire departments. This was understandable. In the final analysis, it was they who had to dig in their pockets and pay the bill for the water damage. Consequently, in order to protect their insured property, many of the larger insurance companies dispatched salvage companies to all fire alarms. The objective of these salvage companies was to protect the building and contents from water damage.

Fire officials' attention was drawn to salvage operations with the development of the Underwriters' Grading Schedule, following the Baltimore Conflagration of 1904. The schedule included the requirement that fire apparatus carry salvage covers. Failure to do so would possibly result in higher insurance rates for the citizens of individual cities. Essentially, this could cause fire chiefs to be blamed for higher rates if they failed to comply with the requirements.

Today, no one seems to argue with the premise that the objective of a fire department is to extinguish a fire with a minimum of loss to life and property. Water damage is part of that loss. In fact, in some fires the water damage is nine to ten times greater than the direct damage caused by the fire. It would seem to follow that salvage operations are a primary consideration of most fire officials. Unfortunately, this is not so. In fact, in some fire departments, salvage operations are almost nonexistent. There are several reasons for this.

One reason that salvage operations are the most neglected of those functions for which a fire department is responsible is that members of most departments are not totally salvage oriented. Most firefighters prefer throwing water or cutting a hole in a roof to spreading sawdust or covering furniture. Consequently, when they become officers they generally give little thought to salvage operations when they are in command of a company or in command of the fire. To overcome this deficiency, it is necessary that all officers be consciously aware of salvage and that salvage operations be so instilled in the minds of all new members during their rookie training that they carry the thinking into fire situations with them.

Another reason that salvage operations are neglected is that very few departments have separate salvage companies. This is not necessarily by choice but primarily a matter of economics. Unfortunately, fire departments throughout the country have been required over the years to do more and more with fewer firefighters and a reduced budget. Most of those departments that had separate salvage companies have had to relegate this function to other companies. However, for the purpose of illustrating the importance of salvage operations in the overall coordination of fire operations, in this book it will be considered that the company performing salvage operations is a separate company.

In most departments, the responsibility for salvage operations is relegated to the truck company. Unfortunately, truck companies in most cities are understaffed. When salvage operations are relegated to understaffed companies that already have the responsibility for laddering, overhaul, ventilation, forcible entry, rescue, and controlling the utilities, something has to give. At some large fires, all of the functions for which the truck company is responsible, with the exception of overhaul, should probably be performed at the same time. It becomes an impossible task, with salvage operations normally being the most neglected factor.

(Incidentally, in those cities where the salvage operations are assumed by the truck company, the word for remembering the functions for which the truck company is responsible changes from *LOUVER* to *LOUVERS.*

There is a way, however, of overcoming this deficiency. If all members of a department are salvage oriented and thoroughly trained in salvage operations, then the officer-in-charge of a fire can call for an extra engine company and assign it the task of performing salvage operations only. Of course, if this type of operation is planned or anticipated, then it is important that all apparatus in the department carry a minimum number of salvage covers and some of the other tools used for salvage operations. It is also good practice for the department to maintain an apparatus in reserve, equipped with a minimum of 20 covers and salvage equipment, that can be dispatched on special call. Two benefits would be derived from this action:

1. There would be a considerable reduction in the fire loss.
2. The public relations of the department would be enhanced.

It is probably worthwhile to expound on item number 2. More letters of appreciation are received by fire departments who are salvage oriented and perform good salvage work at a fire than are received for all the other combined functions performed by departments at a fire. An attic fire in a single-family dwelling serves as a good example. The occupants see heavy black smoke rising from all parts of the house, they see fire coming through the roof, and they watch the firefighters throw water into the building from all sides. They vision all their furniture destroyed, their wall-to-wall carpeting ruined, and their house in shambles. It is not difficult to imagine their surprise and appreciation when they find the opposite. Everything in the house has been protected from water damage. The inside has been cleaned and all the furniture is back in place.

As previously mentioned, this chapter will discuss salvage operations as if a department operated with separate salvage companies under the command of effective salvage officers. The principles of salvage operations are the same, regardless of whether separate salvage companies are used or the truck company is given the responsibility for salvage operations. The big difference is that where separate salvage companies are utilized, good salvage operations generally prevail at a fire.

Salvage work should be started immediately upon the department's arrival at the fire if loss from water damage is to be held to a minimum. If a department is salvage oriented, some of the most effective salvage work can be accomplished through the judicial use of water by engine companies. The problems of the salvage company are reduced considerably when the use of water is held to a minimum. The three operations that should be performed by the salvage company early during the fire operations are (1) spreading covers, (2) bagging floors, and (3) diverting and removing water. They are not necessarily performed in that order. (See Figure 5.1.)

It is important that the salvage officer make a size-up of the problems facing him or her prior to the commencement of operations. The salvage officer should first determine the seat of the fire and make an estimate of the amount of water that will be used for extinguishment and where the excess water will most likely flow. With this knowledge, he or she will be able to establish a plan of action and estimate the number of people it will take to carry out the plan. Additional engine companies should be immediately requested for salvage operations if sufficient personnel are not available on the scene to carry out the salvage plan effectively.

Estimating the amount of water that will be used will help the salvage officer decide on whether or not the water can be held to the fire floor, or whether or not it will be necessary to start operations on the floor below the fire and perhaps all

Figure 5.1 These operations should be performed early during the fire

floors below the fire floor. The type of construction and the type of occupancy will also play an important part in the officer's size-up.

At most fires, salvage work begins and ends on the fire floor. A minimum amount of water is used on the fire and operations are limited to picking up the water with sawdust or possibly diverting it from the building. Sawdust is very effective in picking up water. One bag of sawdust properly used and promptly removed can absorb approximately 30 gallons of water. The sawdust should be spread immediately when the occupancy has wooden floors. This is particularly true if the floor is made of hardwood. Hardwood floors warp easily and are expensive to replace. The sawdust should be picked up as soon as it is spread and the water is absorbed; otherwise, water will start seeping out of the sawdust and back onto the floor.

Sawdust can also be used effectively for damming purposes or to form pathways for pushing out the water. One bag of sawdust is normally sufficient for absorbing and damming the water in a single room of a dwelling or apartment.

There is another important operation that should be performed by the salvage company prior to lines being taken into the building if the fire is on the first floor of a dwelling that has wall-to-wall carpeting, Where possible, a floor runner should be laid from the entrance door to the room on fire. This will provide protection for the carpeting as the firefighters bring in lines and advance them to the fire area. This is particularly important if there happens to be mud or other elements outside that would be taken into the house and over the wall-to-wall carpeting on the boots of the firefighters.

Although it may appear that the water will be held to the fire floor, it is important that the salvage officer immediately survey the floor below the fire to see if there is any possibility that the water will seep through the floor. It may be necessary to start operations simultaneously on both the fire floor and the floor below the fire.

Prior to starting the spreading of covers on the floor below the fire, an estimate should be made of where the water will most likely first come through the ceiling. A logical spot is directly below the fire and through light fixtures in the ceiling.

If the attic or the fire floor is well involved with fire upon arrival, salvage work will normally begin on the floor below the fire. Time is important. A good wooden floor and a plastered ceiling will hold water from 10 to 15 minutes. This will normally provide adequate time to cover items directly under the fire if salvage operations are started early.

In dwelling fires, the furniture should be collected in compact piles. (See Figure 5.2.) It is best that the piles be made away from light fixtures or other ceiling openings, as this will provide more time to do the necessary work. Each group of furniture should be held to a size that can be conveniently covered by a single salvage cover. In a single-family dwelling, one cover will generally cover all the objects in a normal-sized room.

Prior to covering the furniture with a salvage cover, all clothing should be removed from closets and placed in the pile. Additionally, pictures should be removed from the walls and costly draperies and curtains removed from their hangers. Anything of value laying around the room should, if possible, be placed in the

Figure 5.2 Collect furniture in a central pile and then cover it

furniture drawers. And last, any throw carpets on the floor should be rolled and placed on the top of the pile.

Once a pile has been completed and covered, it is good practice to place sawdust completely around the pile where the cover meets the floor. This will prevent water that may get on the floor from seeping under the cover and damaging the legs of the furniture.

When preparing piles for covering, particular care should be taken to protect fragile articles. It is a good idea to protect these articles by covering them with clothing prior to covering the pile with a salvage cover. Articles of clothing placed between these valuables may also help protect them. The purpose of the salvage work could be rendered useless if valuables are broken during the process.

More and more homes have wall-to-wall carpeting. The best way to protect this carpeting is to provide a method of diverting the water to the outside or capturing it in a catch-all as it comes through the ceiling. Damp spots in the ceiling will provide a clue as to where the water is collecting overhead. There are two good methods of diverting the water to the outside: (1) make a drain using pike poles and a salvage cover or (2) make a chute using a ladder and a salvage cover.

To make a drain using a salvage cover and pike poles, lay the salvage cover out flat with the waterproof side up. Place a pike pole on each side of the cover and stick the end of each pike pole through a grommet. Tie it in place if the grommets are too small for the point of the pole. Fold the cover down until it clears the hooks and then roll each pike pole toward the center of the cover until the desired width is reached. Extend the handle end of the improvised drain over the window sill and raise the hooked ends higher than the window sill. Place them on an improvised step ladder or some other available support and secure them in place.

A ladder drain also can be quickly improvised. (See Figure 5.3.) A ladder drain has an advantage in some instances because it can be longer than the length of a salvage cover. Place the end of the ladder over the window sill and raise the other end higher than the window sill and place it on an improvised step ladder or some other suitable object. Lay a salvage cover in the accordion fold on the end of the

Figure 5.3 A ladder drain

ladder that is resting on the window sill. Unfold the cover as it is spread upward on the ladder. If the cover is too short to reach the desired location, overlap the first cover approximately two feet with a second cover and continue to extend the cover up the ladder. Once the pike pole chute or the ladder chute has been completed, use a pike pole or other object to poke a hole in the ceiling where the water is collecting. This will drain the water from the ceiling and allow it to flow to the outside of the building.

It is not always possible to set up a drain to divert the water to the outside because of the distance from where the water will come through the ceiling and a window. In this case it is best to make a catch-all. (See Figure 5.4.) Catch-alls can be made by stretching a salvage cover over chairs, benches, or other small pieces of furniture in such a manner that the cover will bag to the floor. This operation can be performed quickly and will provide for the containment of a considerable amount of water. As with drains, once the catch-all is in place, a hole should be

Figure 5.4 A catch-all made from a salvage cover

poked through the ceiling to allow the water to escape. The sooner the trapped water can be diverted to a drain or a catch-all, the less it will spread out between floors.

In property other than residential, a priority system may have to be established for the order in which material is covered. Early consideration should be given to computers and other sensitive electrical equipment that can be severely damaged by water. A general rule is that stock should be covered before machinery. The reason for this is that machinery is generally less susceptible to water damage. Little damage will occur to most machinery if it is wiped of all water as soon as possible. Of course, an exception to this is machinery that has to be completely dismantled and cleaned if it becomes wet.

The most valuable stock or that which is most susceptible to water damage should be covered first. Particular care should be given to stock in cardboard cartons. The cartons will generally collapse if they are wet, leaving the contents in one big pile. If cartons are stored on the floor, and not skidded, it is best to lift them off the floor and place them on machinery, tables, shelves, boxes, and so on prior to covering. If this is not done, water could get on the floor, under the covered pile, and collapse the entire pile that was covered.

When covering large areas or extensive piles of stock it may be necessary to use more than one cover per pile. This requires that the covers be laid in such a manner as to provide an overlap. An overlap of two feet is generally sufficient to prevent water from getting to the stock. If a large amount of water is expected, it may be best to seal the overlap. This can be done by rolling together the edges of adjoining covers. Lay back the first cover approximately one foot. Place the second cover to where the first is laid back and roll the two together. If a third cover is required, repeat the process where the second and third covers meet.

Belt-driven machinery will occasionally present a problem in regard to covering. If possible, remove the belts. If the belts cannot be removed, cut them at the lacings. This will help solve the problem of covering and little damage will be done to the belts.

Many times vertical piping that extends from floor to floor will provide a pathway for water. The best method of alleviating this problem is to tie a salvage cover around the pipe and form a bag to catch the water.

At a fast-moving fire where a large amount of water is used, it may not be possible to get everything on the floor below the fire covered prior to water coming through the ceiling. The salvage officer should personally supervise operations when faced with this problem. He or she should direct crews on what to cover and what to leave. It is best that firefighters work in pairs. Machinery and stock can be covered much quicker than if a single firefighter works alone. If sufficient personnel is available, a couple of firefighters should be detailed to bring additional covers to the needed area. This works better than having the firefighters who are doing the covering return to the apparatus for additional covers.

If it becomes necessary to bag the floor, it should be done after all covering has been completed. It will generally take a large number of covers to bag a floor in industrial and commercial buildings. The covers should be laid side by side and end to end, with the edges overlapped and rolled to make a watertight seal. This will create a shallow basin that will hold a small amount of water. If more water than can be held by the covers is anticipated, then some means should be made for getting rid of the excess water. The simplest method is to use a shallow-draft electrical pump or some other mechanical device. If none are available, it can be bucketed out or a garden hose siphon can be employed.

Stock is piled in rows with aisles between the rows in many large department stores, mercantile occupancies, and storage occupancies. Protection for this type of arrangement requires the combination of covering stock and bagging the floor. The floor should be bagged after the stock is covered. In these instances, the covers used for bagging should extend part way up over the covers used for covering. This will prevent water projecting from the aisles to the covered stock.

Occasionally it may be difficult to hold the water to the floor below the fire. In these cases it may be necessary to repeat the operation at the next lower floor. In some situations it may be necessary to cover *all* floors from the floor below the fire to the basement. Covering operations must stay ahead of the water flow to be successful. It is better practice to cover a floor unnecessarily than to not cover a floor that should have been.

WATER REMOVAL

It is possible that a building can become so overloaded with water that it will cause the collapse of floors or walls. This is most likely to happen at large fires where a number of heavy streams are projected into the building, at fires in sprinklered buildings where the fire has burned for some time undetected, from broken sprinkler systems, or from broken water pipes in buildings. Every effort must be made to prevent this from happening.

It is generally necessary to remove the water quickly and successfully as soon as possible to prevent the floors from becoming overloaded. There are a number of methods that can be used to achieve this.

One method is to channel the water into permanently installed floor drains or scuppers. (See Figure 5.5.) The water can be channeled by using sawdust or salvage covers. The tools best used for moving the water are squeegees and corn brooms. When using this method to remove water, screens should be placed over the drains or scuppers and should be checked periodically to make sure they are not clogged. The scuppers will normally divert the water directly to the outside of the building, and the floor drains will normally divert the water directly to the sewer system. The floor below, however, should be checked to make sure there is not a leak in the drain system at this point.

Stairways can be used effectively in multiple-story buildings for removing water. The task is simplified if the stairs are concrete with no covering. However, if they are wooden or carpeted it is best that they be covered with salvage covers prior to using them for water removal. Covers used for this operation should be opened up to half their width. It is best to start covering the stairway at the bottom floor. Throw a cover from the bottom step toward the top step as far up as possible. Extend the cover over the banisters on both sides if the stairway is so equipped. The second cover should overlap the first by approximately two feet. Place the cover under the lip of the top step and secure in place. The cover can be secured in place

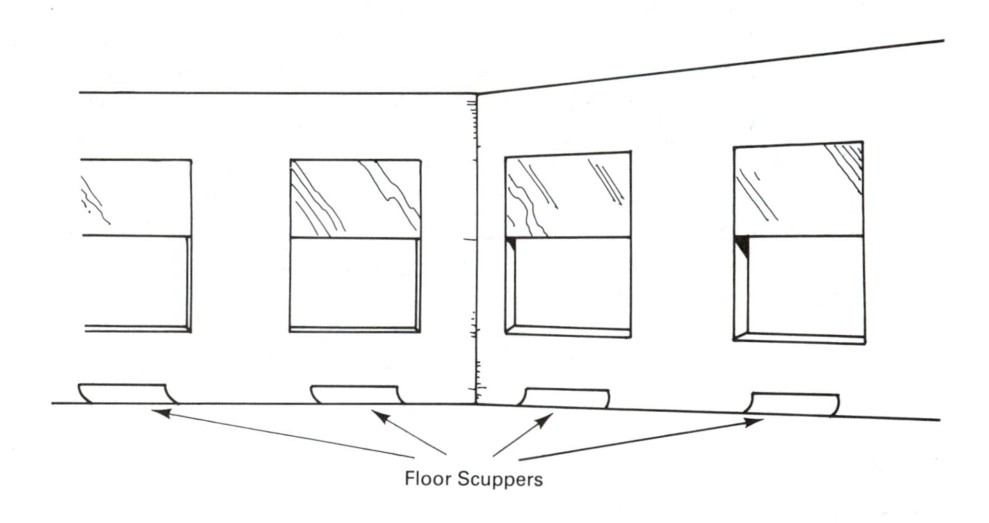

Figure 5.5 Floor scuppers and drains

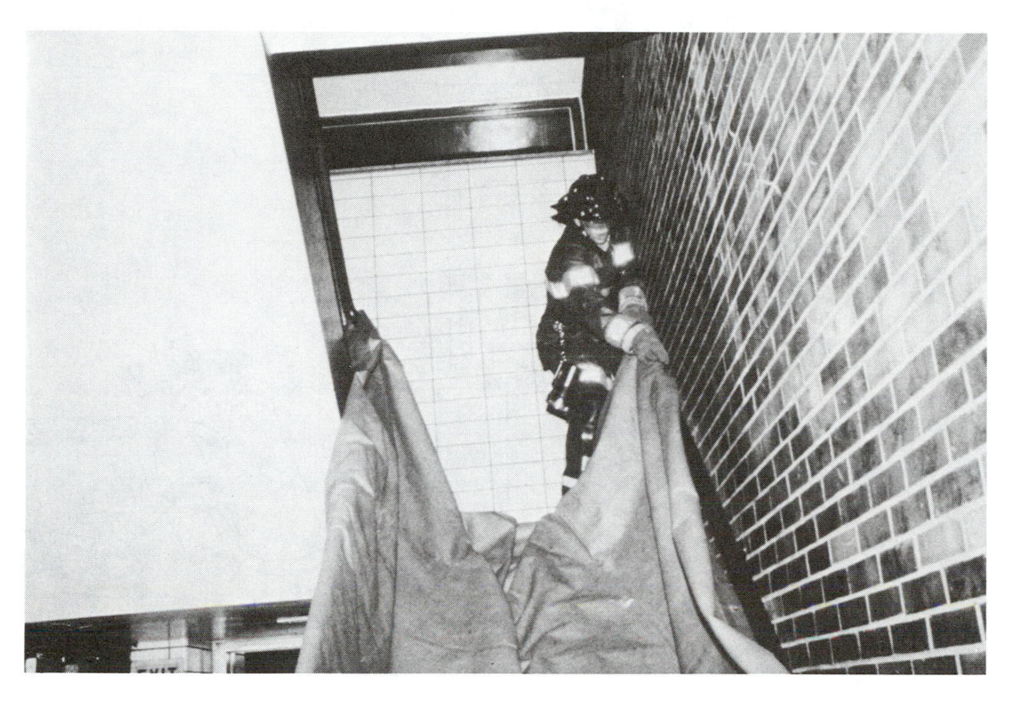

Photo 5.1 Preparing a stairway chute. (Courtesy of Chicago Fire Department)

with a redwood lath, if a bundle is carried on the apparatus. This operation should be repeated from floor to floor.

Sawdust or salvage covers should be used to control the flow of water when stairways are used. Channels should be made to divert the water to the stairway, and also to form dams on each landing to keep the water from being diverted to individual floors.

Another way of clearing water from a floor is by the removal of toilets. (See Figure 5.6.) Once a toilet is removed, a three-or four-inch opening is provided directly to the sewer system. Before removing the toilet, the valve at the bottom of the tank must be shut off if it is the tank type. It is necessary to disconnect the piping on the dry side of the flush valve on other types. Sawdust or salvage covers can be used to channel the water to the floor opening. If possible, some type of screen should be placed over the opening to prevent clogging.

Cast-iron sewer pipes run both vertically and horizontally through buildings. This piping can sometimes be used for water removal. It makes the task easier if cleanouts are available at floor level. If the cleanouts are above floor level and not usable, then it may be necessary to break the pipe. It may also be necessary to cut into a wall or floor to gain access to the pipe.

If there is no other practical means of removing water from a floor, then it may be necessary to cut a hole in the floor. The hole should be cut close to an outside window and be of sufficient size to handle a large amount of water. All stock and other material located below the spot where the hole will be cut should be removed prior to performing this operation. A drain should be set up at this

Figure 5.6 Removal of a toilet provides a direct line to the sewer system

point that will divert the water to the outside of the building. A ladder and salvage cover can be used for this purpose. Once the hole is cut, it is best to enlarge it from below so that water will not be diverted to the area between floors.

It may be necessary to breach a wall in those instances where a large amount of water has collected on a floor and there are a limited number of methods for removing it. This operation can best be performed by selecting an outside window that has a window sill relatively close to the floor. It is necessary to remove the glass and window frame. The wall below the window sill can then be removed. This provides a large, clean opening directly to the outside.

Elevator shafts can also be used for water removal but should be low on the priority list. (See Figure 5.7.) If the shaft is to be used, the cage should be moved to a floor above the upper floor from which water will be removed and the elevator taken out of service. It will be necessary to pry open the door at the upper floor and also at the first floor. If the elevator machinery is located at the bottom of the shaft, it should be covered before water enters the shaft. As early as possible, a siphon should be set up at the bottom of the shaft to remove the water from this location to the outside of the building. If the building has a basement, a check should be made to ensure that the water is not leaking out of the shaft and into the basement area.

FIRES IN SPRINKLERED BUILDINGS

Fires in sprinklered buildings almost always present a problem to salvage company officers. (See Figure 5.8.) Although the large majority of all fires in sprinklered

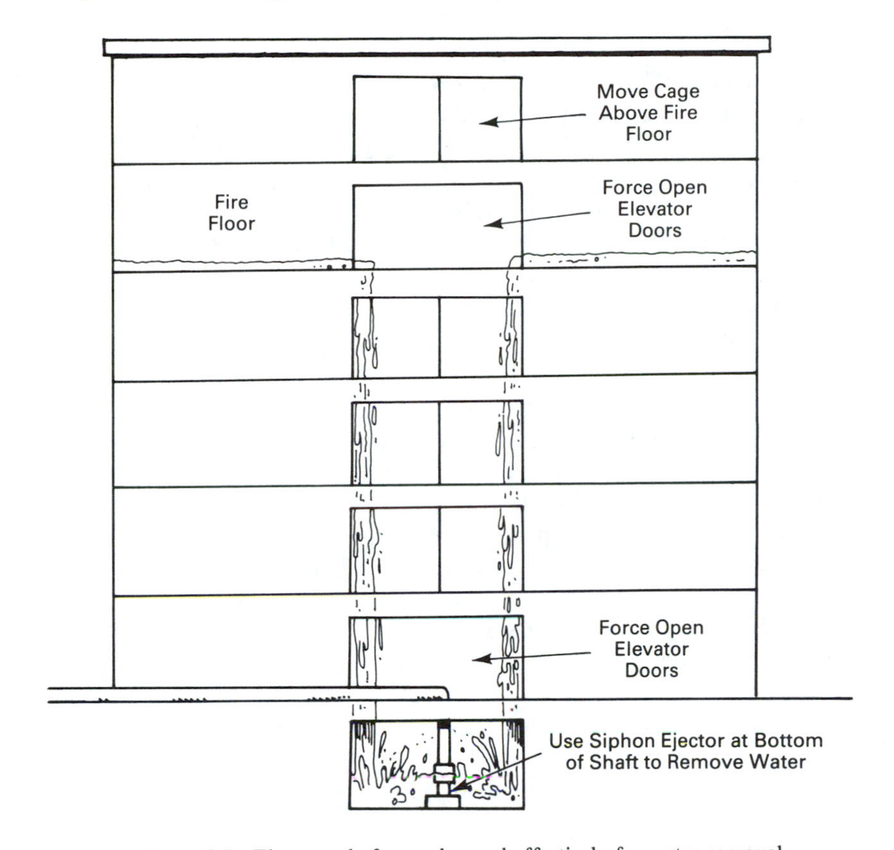

Figure 5.7 Elevator shafts can be used effectively for water removal

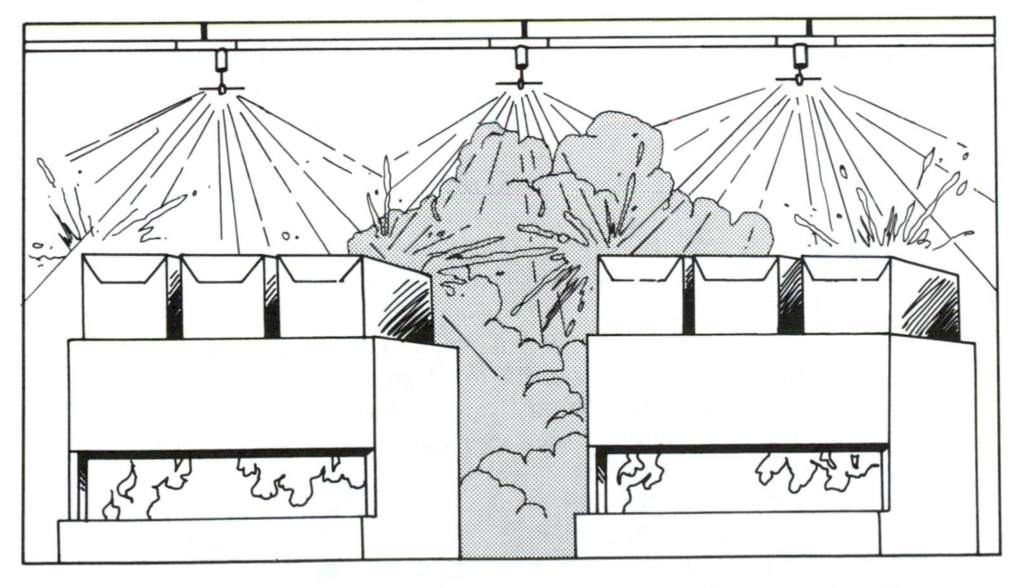

Figure 5.8 Fires in sprinklered buildings create a water control and removal problem

buildings are extinguished or held in check by the sprinkler system, water continues to flow until the system is shut off. Sprinkler system control, in such cases, is normally the responsibility of the salvage company officer. He or she must, however, keep in mind that the purpose of the sprinkler system is fire control and that a complete shutdown of the system should never be made until there is absolute assurance that the fire has been extinguished or that hose lines have been laid and are in position to ensure that there is no danger of fire spread. Prior to shutting down the system, salvage operations will consist of spreading covers and controlling water being discharged by opened heads.

A check should be made of adjacent buildings if the fire is of sufficient size to cause an exposure problem. It is possible for the heat from a fire to enter an adjacent building and activate one or more sprinkler heads. To prevent this, close the windows on the exposed side and open them on the unexposed side. This will result in restricting the entrance of additional heat into the building and reduce the internal heat caused by the fire. If sprinkler heads have been activated, shutoffs should be placed in the heads; however, the system should not be shut down until the exposure problem has been eliminated. It will be necessary to put this system back into operation prior to leaving the scene of the emergency.

Salvage companies should carry devices for shutting off the flow to individual activated heads. These are normally inserted into the heads prior to the system being shut down. Most of the sprinkler shutoffs available are capable of completely shutting off the flow from a head, as long as the head is intact. Some departments use manufactured sprinkler shutoffs for this purpose; other departments use wooden wedges or doorstops. Occasionally, however, a head may be broken off in such a manner as to prohibit shutting off the flow of water at the head. Additionally with some flush-type heads it is difficult to stop the flow of water at the head by using available shutoff devices. A possible solution to this problem is to improvise a drain from the individual head. A 4-to 2½-inch reducing fitting can be attached to a section of 2½-inch hose and the hose extended to the outside of the building. The reducing fitting can then be secured over the ruptured head and held in place. This will divert the flow to the outside of the building. It will generally be necessary to hold the reducer in place until the system has been shut down and drained. (See Figure 5.9.)

Once the salvage officer is assured that the fire is out or adequately controlled, the sprinkler system may be shut off. However, prior to doing so the salvage officer should notify the officer-in-charge. He or she should also notify the officer-in-charge when the system is put back into operation.

It will be necessary for the salvage officer to locate the main sprinkler shutoff valve to shut down the system completely. This valve can normally be quickly located by noting the location of the fire department steamer connection on the outside of the building. The shutoff valve is usually inside of the building relatively close to the point where the steamer piping enters the building from the outside.

It is the practice of most departments for the first- or second-in engine company to lay into the sprinkler steamer connection whenever there is any possibility of a fire in a sprinklered building. Generally the pump operator will start flowing water to the system if there is any indication whatsoever that there is a fire in the

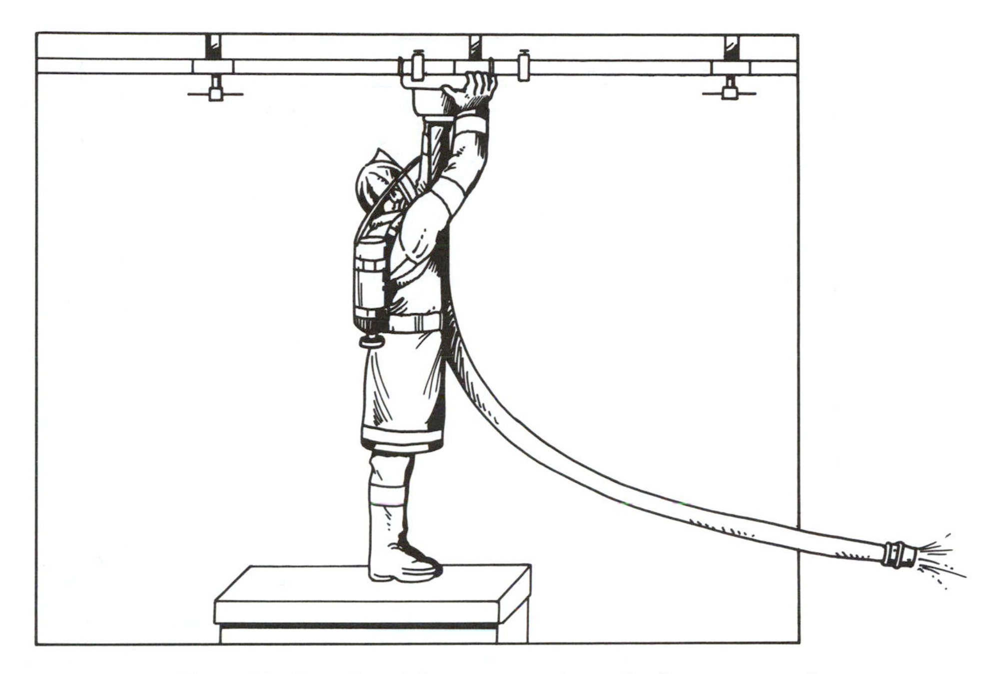

Figure 5.9 Sometimes it is necessary to improvise for water control

building. The fire department steamer connection bypasses the main shutoff valve; consequently, water will flow into the system from a pumper even if the main shutoff valve is closed. (See Figure 5.10.) It is therefore important that the salvage officer be sure that no pumper is pumping into the system prior to his closing the main shutoff valve.

The sprinkler system can be shut down by first closing the main shutoff valve and then opening the drain valve. Most systems have a test valve at the far end of the system. This valve should be located and opened, and a firefighter should be stationed at the location. Opening this valve will allow air to enter the system and provide for a better drainage of the system. The ruptured heads can be removed and replaced once the system is drained.

In some communities the replacement of heads is the responsibility of the owner or sprinkler company. In other communities the fire department is responsible for replacing the heads and leaving the system in working order prior to leaving the premises. When the fire department is responsible for this function, there normally will be an adequate supply of heads available on the premises to replace the ruptured heads. If not, most salvage and truck companies carry a sufficient assortment of heads to make the necessary replacements. There are several points that should be remembered in replacing the heads:

1. Replacement heads should be of the proper temperature rating. The temperature rating of solder-type sprinklers is stamped on the solder link. For other types of heads it is stamped on one of the releasing parts. If the releasing parts cannot be

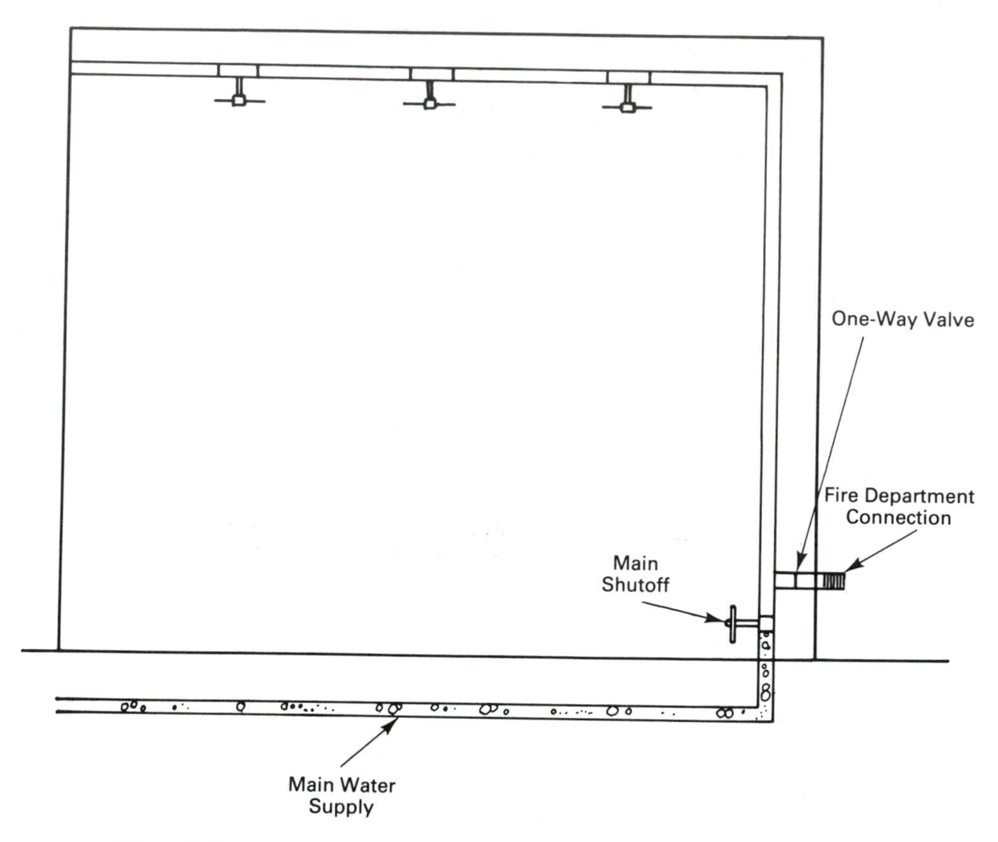

Figure 5.10 The fire department connection bypasses the main sprinkler shutoff

found, it may be necessary to check the rating of one of the adjacent heads that has not been activated to determine the proper temperature rating for replacement.

2. Upright sprinkler heads should be installed with their frame work parallel with the pipe so as to not disturb the spray pattern.

3. Upright and pendant-type heads should be replaced with heads of the same type.

The system is ready to be put back in service once the ruptured heads have been replaced. First close the drain valve and then open the main shutoff valve. Water will be discharged from the test valve once the system is nearly filled. This valve should be closed when a steady flow of water is being discharged.

POSTFIRE SALVAGE OPERATIONS

After the main body of fire has been extinguished, salvage operations will consist of doing whatever is necessary to prevent additional water damage. The work includes removing water, as previously discussed; drying machinery, stock, furniture, and so on; removing salvageable articles from the debris; and assisting the truck company in covering the roof, if necessary.

Some good salvage work can be done with the use of a sponge and chamois. These can be used to dry furniture and fixtures and thereby prevent the finish from being water spotted and ruined. They can also be used to dry machinery. Where possible, the machinery should be oiled after being dried to prevent rusting.

It will also be necessary after the fire to remove all salvage covers. They should not be removed, however, until all the water has stopped dripping. It may be necessary to leave the covers in place if it appears that dripping will continue for a long period of time. In such instances it is best to secure a signed agreement from the building owner accepting full responsibility for the safe return to the department of all covers.

Some debris and water may have collected on top of some of the covers. The debris should be checked to ensure that it does not contain any hot embers. It may be necessary to wet down the debris prior to its removal. A hot ember can easily burn through a cover, allowing the possibility of igniting the material being protected.

Care should be taken during the removal of a cover to ensure that the contents protected during the fire are not damaged by spilling water or debris on them. A cover should be folded from all sides in such a manner as to contain the water and debris inside. Care should also be taken so as not to damage any of the fragile articles that were protected. Once removed, the cover should be carried outside in such a manner that none of the debris or water trapped inside is spilled during transportation.

Stairway drains should be removed as soon as possible. This should be done so as to protect the covers from people walking over them. Floor runners can be placed over the stairs if it is necessary to protect finishing or carpeting from mud or debris.

Small holes in the roof are generally covered by the truck company if it becomes necessary to protect the interior of the building from rain or other weather elements. If the holes are large, or if the entire roof must be covered, it may be necessary to use salvage covers. This operation is generally a joint effort of the salvage and truck companies. If the entire roof is burned away, it may be necessary to bag the attic, the floor below the attic, or both. Again, if it is necessary to leave salvage covers on the premises, a signed acceptance by the owner for responsibility of the covers should be obtained.

REVIEW QUESTIONS

1. What was the objective of insurance companies' salvage companies?
2. What are two reasons that salvage operations are almost nonexistent in some fire departments?
3. What company normally assumes the responsibility for salvage operations in those departments not having separate salvage companies?

4. What is a method that can be used to overcome the deficiency of not having separate salvage companies?
5. What are two benefits of doing effective salvage work at a fire?
6. When should salvage operations commence at a fire?
7. How can the engine company be effective in salvage work?
8. What are the three operations that should be performed by the salvage company early during fire operations?
9. What are the first items that a salvage officer should determine in his or her size-up of the fire?
10. What should the salvage officer do if there is an insufficient number of personnel on the scene to carry out his or her plan effectively?
11. At most fires, where does salvage work begin and end?
12. Approximately how much water can one bag of sawdust absorb if properly used and promptly removed?
13. Why is it important to pick up sawdust as soon as it is spread and the water absorbed?
14. How much sawdust will it normally take for absorbing the water and damming operations in a single room of a dwelling or apartment?
15. What is an important operation the salvage company should perform prior to lines being taken into the building if the fire is on the first floor of a dwelling that has wall-to-wall carpeting?
16. What should the salvage officer do prior to starting to spread covers on the floor below the fire?
17. Where are the logical spots that water will come through the ceiling of the floor below the fire?
18. Approximately how long will a good wooden floor and a plastered ceiling hold water?
19. How many covers are normally needed to cover the furniture in a normal-sized room of a single-family dwelling?
20. In a dwelling, what should be put in the pile to be covered?
21. What should be done once a pile has been completed and covered in a dwelling?
22. How can fragile articles be protected when placing them in the pile in preparation for covering?
23. What is the best method of protecting wall-to-wall carpeting in a room from water from the floor above?
24. What are two good methods of diverting water from the floor above to the outside as it comes through the ceiling?
25. How is a pike pole drain made?
26. How is a ladder drain made?
27. What is the next best thing that can be done if it is not possible to make a pike pole or ladder drain to catch the water from the floor above.
28. How is a catch-all made?
29. In general, in what priority order should the following items be covered: machinery, stock, sensitive electrical equipment?
30. What should generally be used as a guide for setting up a priority list for covering?
31. What should be done with filled cardboard cartons if they are stored on the floor and not skidded?

32. What should be done when it is necessary to use more than one salvage cover to cover a large area or extensive pile of stock?
33. What should be done with belt-driven machinery if the belts hamper the covering process?
34. What should be done to alleviate the problem of water coming through holes made for vertical piping?
35. What is the procedure used to bag a floor?
36. What is the simplest method of getting rid of the excess water when more water comes through the floor than can be held by bagging?
37. What method should be used to protect the stock when it is piled in rows with aisles between the rows?
38. What are some of the situations where so much water can be collected in a building that it will cause the collapse of floors or walls?
39. What are some of the different methods that can be used to remove water from a building?
40. What method should be used to protect the steps if the stairway is to be used for water removal?
41. How should the water be shut off if a toilet is to be used for water removal?
42. What must be done if an elevator shaft is to be used for water removal?
43. What must be done before a sprinkler system is shut off?
44. What precautionary measures should be taken in sprinklered adjacent buildings that are exposed to a large fire?
45. How can water be diverted from a broken sprinkler head if it is going to take some time to shut down the system?
46. What should the salvage officer do prior to shutting down a sprinkler system?
47. What is a good method of locating the main sprinkler shutoff valve?
48. What should the salvage officer make sure of regarding the fire department steamer connection prior to shutting off a sprinkler system?
49. What is the procedure for shutting down a sprinkler system?
50. What are several things to remember regarding the replacement of sprinkler heads?
51. What procedure should be used for putting a sprinkler system back into operation?
52. What are some of the things that should be done in performing postfire salvage operations?
53. How should salvage covers be removed from a pile?

CHAPTER 6

Basic Firefighting Tactics

OBJECTIVE

The objective of this chapter is to explore the basic firefighting tactics involved in locating, confining, and extinguishing a fire. The basic procedures used in making a direct or indirect attack on a fire are examined. Review questions are provided at the end of the chapter to assist the reader in determining his or her level of comprehension of the chapter contents.

Firefighting strategy and firefighting tactics are two terms commonly used when discussing fireground operations. There is a difference in the two but the difference is small. Firefighting *strategy* is best related to the planning of operations, whereas firefighting *tactics* are related to the carrying out of the plan. For example, the officer-in-charge decides that a hole should be cut in the roof for ventilation. This is part of the officer's strategy for controlling and extinguishing the fire. He or she gives an order to the captain of the truck company to cut the hole. The size of the hole, as well as where and how it will be cut, are left up to the truck company officer. His or her carrying out of the plan to cut the hole in the roof is tactics.

Although a number of changes have occurred in firefighting apparatus, tools, and equipment over the years, very few changes have been made in the basic principles of firefighting strategy and tactics. Furthermore, as we look toward the future we can visualize that technological advancements will have their impact on fire apparatus, tools, and equipment. Additionally, the Incident Command System will have its impact on strategy and tactics as applied to command and control; however, the basic principles of firefighting will probably remain nearly the same. Consequently, anyone entering the fire service should learn these basic principles well, as he or she will probably be living with them throughout an entire career.

In its simplest sense, basic firefighting strategy is the making of plans for (1) locating the fire, (2) confining the fire, and (3) extinguishing the fire. It therefore follows that basic firefighting tactics is the carrying out of this plan.

LOCATING THE FIRE

When most people think of a fire, they envision lots of flames and smoke. It is not always exactly like that, however. Furthermore, from a firefighting standpoint, locating the fire encompasses more than just searching for and finding the flame. Tactically, it includes (1) locating the seat of the fire, (2) determining what is burning, (3) determining the extent of the main body of fire, and (4) evaluating where and how fast it is likely to travel.

Locating the Seat of the Fire

The seat of the fire is the hottest part of the fire. It is usually located at the point of origin, but not always. Since the seat of the fire is producing heat faster than at any other location, maximum return on the water used can normally be obtained if water is directed at this point. However, it is of course first necessary to find and identify this location.

The identification of the seat of the fire is usually not difficult if the so-called burning materials are producing flame and the fire is relatively small. It becomes a problem, however, when flame cannot be found.

There is an old saying that where there is smoke there is fire. Don't you believe it. Fire was previously defined as rapid oxidation accompanied by heat and flame (or heat and light). It is possible to have a smoldering fire with some glowing embers, with no flame but lots of smoke. It is also possible to receive an alarm of fire and upon arrival find a building or portion of a building filled with smoke but no visible signs of a fire. In these cases it is probable that the fire has gone out or there has been a source of heat that has produced a lot of smoke, followed by the elimination of the heat source. Finding the cause of the smoke falls under the category of locating the fire.

The sense of smell will many times provide a clue as to the possible source of the smoke. The alarm classified as "food on the stove" is a typical example. At times the smell is so strong that the first arriving company officer can determine the source of the smoke when the apparatus first pulls up in front of the dwelling.

The sense of smell may also be useful in extremely light smoke conditions. Of course, if there is any amount of smoke whatsoever, breathing apparatus will be worn, which will render useless the sense of smell.

Some burning materials such as cloth, rags, or insulation on electrical wires, also give off a distinct odor. If the smoke is light, using the sense of smell may guide the search party immediately to the source; however, at other times it is extremely difficult to locate the source by using the sense of smell. This is particularly true if the source is not found immediately. Even a small amount of smoke will quickly dull the sense of smell, and the longer a searcher is in an atmosphere containing any amount of smoke, the less he or she can depend on this sense as an aid. When this happens it is best that the searcher return to the outside and take in a few breaths of air and then return for a second try. Better still, if there is difficulty finding the source after a few minutes of search, it might be better to bring in someone who has not previously been exposed to the smoke and give him or her a try.

The appearance of the smoke will sometimes give a clue as to whether the source is still active. Heat will cause the smoke to be buoyant. It will rise and swirl about, the degree of which depends on the amount of heat providing the stimulus. Smoke in this condition is referred to as *live smoke. Dead smoke,* on the other hand, is not buoyant. It will remain relatively still and may collect in pockets. Dead smoke normally indicates that there is no active fire and that the source of the heat has been eliminated.

Regardless, it is necessary to locate the source of the smoke. This can be done only through a systematic search. The search party should be organized and each firefighter given a definite assignment. A group of firefighters wandering aimlessly about normally produces nothing. Check all possible sources. One of the first things to check is electrical wiring, equipment, and appliances. Search out all electrical motors. Remember—motors can be found in refrigerators, air conditioning systems, heating systems, escalators, elevators, washers, dryers, and so on. The best method for a firefighter to determine whether or not a motor could have produced the smoke is to place the back of his or her hand on the motor to see if it is hot. When doing this, the person should make it a rapid touch so as not to burn his or her hand if the motor has been overheated. Carefully check neon lights and fluorescent fixtures. Light ballasts in fluorescent fixtures are a common cause of trouble. Check electric wires that are plugged into outlets to see if there is any smoke around the outlet, and also all electrical appliances for the possibility of recent heat. All of the walls should be observed to see if there is any discoloration of the wallpaper. This would most likely indicate a partition fire. The other walls should be felt with the hand to make sure that there is no fire hidden there. Check all locations, including storage closets and hallways.

Sometimes the source of the smoke may be some distance from the body of smoke itself. A good example is a multiple-story building with the top floor filled with smoke. (See Figure 6.1.) When arriving on the scene where this condition exists, it is good practice to check out the top floor and the basement simultaneously. Fires in basements of multiple-story buildings have a habit of sending large quantities of smoke up unprotected openings to fill the top floor with smoke.

Another type of occupancy where the location of the smoke can be deceiving is a string of mercantile occupancies with a common attic. The determination that the attic is filled with smoke should initiate a search of not only the attic but all the occupancies below the attic.

Do not be too quick to leave the premises if the source of the smoke cannot be found. Make sure that nothing is overlooked. Particular care should be taken in hotels and apartment houses. If necessary, every room in the building should be opened and checked. Sometimes an occupant will have a fire or an extinguished fire in his or her room and try to hide the fact from the firefighters. This is particularly true when there is something in the room that the occupant does not want outsiders to see. The individual will extinguish the fire himself or herself, not call the fire department, and when the firefighters knock on the door respond with the answer, "There's no fire in here." No harm can generally be done to accept this type of

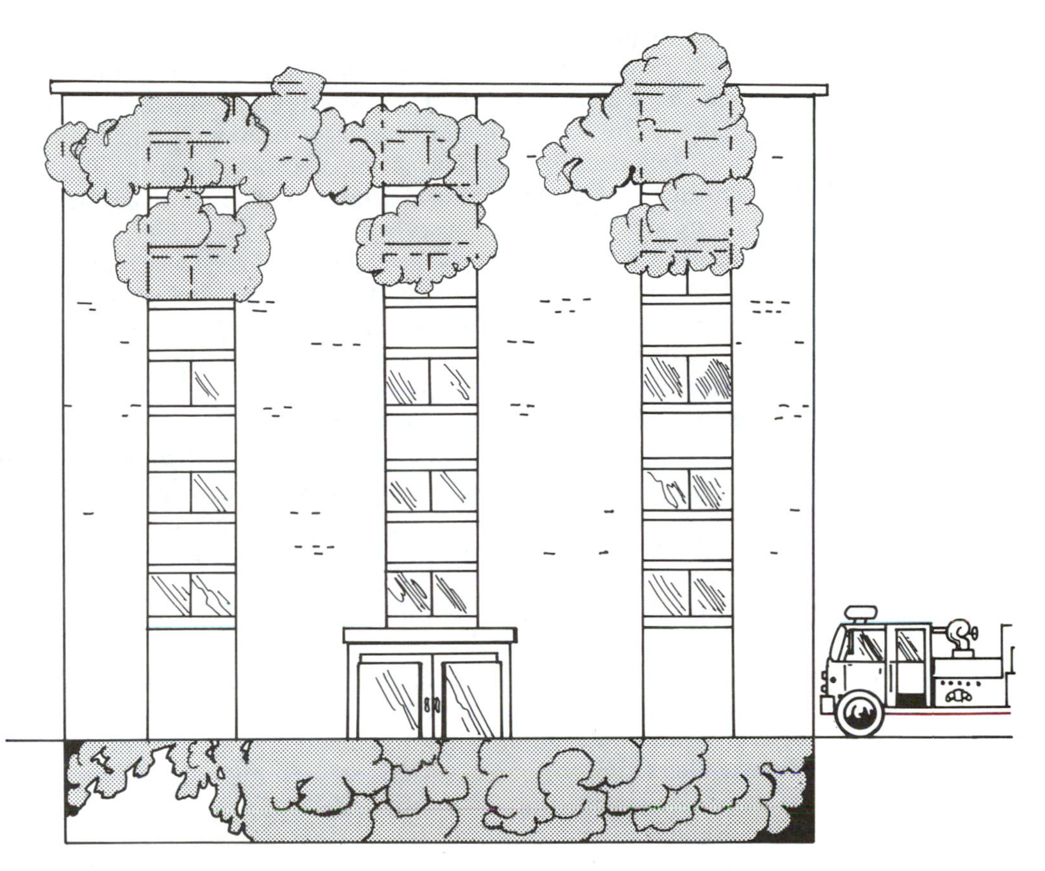

Figure 6.1 Where's the fire?

response when it is first given; however, do not fail to return to each of these locations and demand a "look-see" if the source of the smoke cannot be found elsewhere. Another reason to check every room is that the occupant may have had a fire, put it out, and was overcome by the smoke. There is at least one case on record where this occurred and only a diligent and thorough search by the firefighters saved the person's life.

Determining What Is Burning

Determining the type and amount of material involved in the fire is extremely important in developing the tactics to be used for extinguishment. This knowledge is required for selection of the extinguishing agent and, if water is to be used, the number and size of lines that will be required. However, obtaining this information is not always easy. The task is somewhat simplified if a previous inspection has been made of the building and the responding officers are familiar with what was stored in the area involved. Otherwise, if the fire is of any size, it might be necessary to estimate what is burning by what is seen and the type of occupancy involved. For example, from past experience firefighters know that in dwellings, office buildings, and small mercantile establishments the fire will normally be fed by various types of ordinary combustible materials. However, it is good practice not to become complacent, as the materials involved may not be what is commonly expected in that type of occupancy. A hotter fire than normal or one that is moving faster than expected could indicate that an accelerant had been used to start the fire or that flammable liquids are stored on the premises.

Another clue as to what is burning is the color, amount, density, and odor of the smoke. Materials such as roofing paper, asphalt, rubber, and petroleum products usually produce clouds of heavy black smoke. A white, grayish smoke usually indicates a grass fire. Abnormal colors in the smoke, such as yellow and red, generally indicate that chemicals are involved. A normal room fire containing furnishings and painted walls will generally produce a dark gray smoke, whereas a structure such as a house or garage gives off a medium brown smoke. If the structure and the contents are both well involved, the smoke produced may be almost black. Some products have characteristics all of their own. Magnesium burns with a brilliant white light, and hydrogen burns with no visible flame or smoke.

Determining the Extent of the Fire

After determining what is burning, it is necessary to find out the extent of the fire. Is it confined to a single room or are there several rooms involved? Has the fire spread to adjacent exposures? Is an entire floor of a multiple-story building involved or has the fire extended to upper floors? Answers to these questions will help determine the personnel, equipment, and number and size of lines required for confinement and extinguishment.

When evaluating the extent of the fire it is important that all fire be located. There have been cases where firefighters have shut down their lines in preparation for picking up only to discover that the fire was burning fiercely in another portion of the building. Particular care should be taken in multiple-story buildings to check the floors both below and above the fire floor, and also the basement and top floor. Don't forget the possibility of partition fires, fires in hidden areas, and the potential of fire spread to other portions of the building through air-conditioning ducts or unprotected vertical openings. Keep in mind the thought that the fire could have been started by an arsonist who also set fires in other sections of the building.

Determining Where and How Fast the Fire Will Travel

The construction of the building is the best clue as to where the fire will spread. Thought should be given as to whether or not the building has interior walls and fire doors that will prevent or contribute to the spread of the fire. How about open stairways and other unprotected vertical openings? In addition to hindering or contributing to the spread of the fire, the construction features also play an important part in how fast the fire will travel. The walls and ceilings may act only as a pathway for the fire travel, or they themselves may burn and contribute to the rapid spread of the fire.

When evaluating where and how fast the fire will progress, thought should be given to the four methods by which heat travels from one body to another. Of the four (direct contact, conduction, radiation, and convection), direct contact, radiation, and convection play the biggest part in rapid travel of a fire. Fire will travel horizontally and vertically by any one of these methods, with vertical travel being influenced primarily by convection. Fire will move in every direction until stopped by a barrier and will be pulled toward any opening that provides an outside source of oxygen. Consequently, determining where the fire will go is generally one of locating open pathways for its travel and outside openings that will pull it in that direction.

Making an estimate of where and how fast the fire will travel is necessary in order to put into effect the second stage of the basic firefighting tactics—confining the fire. Consequently, the information should be used to determine at what locations stands will have to be made in order to stop the spread of the fire, and how many lines it will take to accomplish the task. Planning should include conceding some areas that are already lost and some that are not yet burning in order to get ahead of the fire. It will therefore be necessary when determining where and how fast the fire will travel to estimate where the fire will be by the time lines are laid and moved into position. Thought should be given to how much water will be needed to hold or stop the fire and how much is readily available. The basic principle involved is always to make plans to *stay ahead* of the fire. Don't get caught in the trap of chasing it.

CONFINING THE FIRE

From a practical firefighting standpoint, the procedures required to confine a fire will vary with the type of fire. For example, different procedures will be involved for confining a brush fire than for confining one in a truck containing LPG or for a fire in a structure. However, the basic principles involved are similar. Consequently, confining the fire in structures will be discussed in this chapter. The variances as applicable to individual types of fires will be covered in later chapters.

From a theoretical standpoint, confining the fire means to prevent its spread beyond the point of origin. From a practical point this is impossible, unless of course the fire is discovered when ignition first takes place and an extinguishing agent can immediately be applied to it.

From a practical rather than a theoretical standpoint, two different objectives can be established for the tactics used to confine a fire in a structure.

1. Prevent the spread of the fire from the building of origin, or from the buildings involved at the time of arrival of the fire department. Achieving this objective is generally a matter of protecting the external exposures and patrolling downwind from the fire for flying brands. Protecting the external exposures is a matter of keeping the temperature of the exposure below its ignition temperature. This is normally done by wetting down the exposure, setting up a protective curtain between the exposure and the heating source, reducing the amount of heat produced by the source, or a combination of the three.

2. Prevent the interior spread of the fire both horizontally and vertically beyond the point where the fire is met once hose lines have been laid and are in place.

In the development of tactics to achieve these two objectives, it should be kept in mind that at the majority of all fires *both objectives are achieved by a rapid, aggressive attack on the fire.* At the remaining fires, the objectives are met by a combination of aggressively attacking the fire, protecting external exposures, and preventing horizontal and vertical internal spread. In summary, *at all fires, confining the fire requires an aggressive attack on the main body of fire.* In essence, this means that at all fires, one of the most effective methods of protecting the exposures is to reduce the amount of heat produced by the heating source.

Preventing External Spread

Preventing external spread is primarily a matter of protecting the external exposures. (See Figure 6.2.) Following is a typical example of the tactics used to achieve this objective.

The department arrives on the scene to find one or two rooms of a one-story,

Figure 6.2 Protect the exposures

single-family dwelling well involved with fire. Fire is shooting out the windows and the radiant heat is threatening the house next door. To keep the fire loss to a minimum, it is necessary to prevent the exposed house from bursting into flames and the fire from extending to other portions of the involved building. Both objectives can normally be achieved through the use of $1\frac{1}{2}$- or $1\frac{3}{4}$-inch lines. Consequently, the general tactic is for the first arriving engine company to lay a large supply line into the inlet of the pumper and work the smaller lines off the pumper's outlets. One of the $1\frac{1}{2}$- or $1\frac{3}{4}$-inch lines should be advanced to a point where the firefighter can direct water onto the exposed dwelling while the second line is taken inside to prevent further spread and extinguish the fire.

Let's extend the example situation further. The first arriving company finds the dwelling well involved with fire. Fire is going through the roof and extending out windows on all sides. Houses on both sides of the burning building are in immediate danger of being ignited from the radiated heat. In this case, the survival profile indicates that there are probably no survivors and that there is very little in the burning building that can be saved. Consequently, the objective is to prevent the adjacent buildings from becoming involved. Therefore, both of the $1\frac{1}{2}$- or $1\frac{3}{4}$-inch lines are used outside to protect the exposures. (See Figure 6.3.) One is taken in between the fire building and the exposure on one side with water being quickly projected onto the exposure. The other is taken to the opposite side and used in the same manner. The second engine company to arrive is given the responsibility of attacking the fire.

Many times this second type of operation requires that an explanation be given to the owner of the burning dwelling and perhaps also to the neighbors. The general public does not understand the principles involved, and reacts only to what they have seen. They have probably been outside for a few minutes watching the fire get bigger, wondering why the fire department is taking so long to get there. The owners of the burning building are perhaps in tears, watching all they have worked for going up in flames. The fire department arrives on the scene, lays lines, and starts squirting water on the buildings on both sides of the fire—and these buildings aren't even burning. The question immediately arises in the minds of the people watching, "What kind of fire department do we have?" It is much better to educate them now than to do so in the media after the unfavorable publicity has hit the press.

There is very little difference between the tactics used to protect the external exposures at a fire in a dwelling than those used at a fire in an industrial or commercial building. The big difference is in the size of lines required, the number of personnel needed, and the lack of experience of most firefighters with this type of operation. Let's pause for a minute on the last point. In some fire departments, it is possible for a firefighter to spend his or her entire career in the department and not man a $2\frac{1}{2}$-inch hose line or a heavy stream appliance at a fire more than a half dozen times. This is not stated in a derogatory manner but more as a fact. Fortunately, the lack of experience is compensated for in most departments by training. Most of the initial training received by a recruit firefighter at the drill tower is devoted to preparing him or her to operate at a large fire. Furthermore, most of the monthly training in hose lays received at the company level is devoted to this objective. Consequently, although lacking in experience, operations at large fires do not come as a surprise to most firefighters.

Figure 6.3 Use the first lines to protect the exposures

Operations at a larger fire might progress in a manner similar to the following: The officer of the first arriving engine company finds a one-story industrial building partially involved with fire. Fire is shooting out the windows on the north side and threatening an adjacent one-story building. At this point the officer informs the dispatch center of the situation and requests the number of additional companies he or she thinks will be needed for extinguishment and confinement. The officer may then lay a line and make an attack on the fire or use the line to protect the exposure, depending on his or her analyses of the seriousness of the threat to the exposure. If the decision is made to attack the fire, the officer will call for the next arriving company to protect the exposure. It may even be possible that two companies will be needed for exposure protection.

When wetting down an exposure, it is generally best to direct the stream near the top of the wall and allow the water to run down the side. This procedure will keep the entire wall cooled. Of course, if any portion of the exposure is immediately threatened, it should be wetted down thoroughly prior to the above procedure being established.

Photo 6.1 Heavy streams can be used effectively to protect exposures. (Courtesy of Chris Mickal, New Orleans Fire Department Photo Unit)

The difference in protecting this exposure and protecting the exposure at the dwelling is that $1\frac{1}{2}$- or $1\frac{3}{4}$-inch lines could be used at the dwelling, whereas $2\frac{1}{2}$-inch lines or heavy stream appliances will be needed at the industrial site. This means that a company that was capable of manning two lines at the dwelling fire will be restricted to handling one at the larger fire. In many cases the immediate threat to the exposed dwelling is more serious than the immediate threat at an industrial fire. The reason is that the exposure is closer to the source of heat at the dwelling fire and the exposure is generally made of material that will burn easier. However, at the industrial fire the exposed area is generally larger and the heat being given off by the source is more intense. In both cases the need and tactics are basically the same. Lay the line or lines required and keep the exposure wetted down until the threat of ignition has been eliminated.

Let's extend this fire even further. The first arriving officer finds the building well involved with fire and flames shooting out all sides and through the roof. There are exposures on all sides. The officer calls for the help he or she feels is needed and starts directing companies. Protection of the exposures is achieved in this situation in the same manner as in the previous illustration. The difference is that it will take more lines and more personnel, and more decisions will have to be made as to the priority for the placement of lines.

The priority for placement of lines for exposure protection should be based on the order in which the exposures will become involved with fire if something is not done. Generally, the exposure on the leeward side of the fire will receive first priority; however, the distance between buildings, the construction of the exposures, and the intensity of the fire at various locations will all influence the priority arrangement. It should also be mentioned that with this type of fire, it is good practice to have an engine company patrolling downwind as a precautionary measure against flying brands. As a further precautionary measure, it is good practice to set up for the protection of buildings that are not exposures upon arrival of the department but may become exposure problems before the fire can be successfully contained.

The protection of exposures from fires in multiple-story buildings presents a somewhat different problem. Fortunately, as buildings rise in height, the distance between buildings generally increases and construction improves. The improvements, however, do not eliminate the exposure problem. Fire has demonstrated the ability to cross wide streets and ignite buildings on the other side. A prime example is the Burlington Building fire that occurred in Chicago a number of years ago. At this fire, radiated heat from a group of burning buildings carried across an 80-foot street. The result was the ignition of the upper 7 floors of a 15-story office building. Radiated heat was estimated to have reached a temperature of 1,800 degrees Fahrenheit.

It is possible to protect the exposures to a multiple-story building fire by wetting them down if the exposures are within the reach of streams from water towers, ladder pipes, elevated platforms, or other heavy stream appliances. The streams should be directed well above the heated area to allow the water to cascade down

Photo 6.2 Narrow streets increase the exposure problem. (Courtesy of Chicago Fire Department)

the wall and protect as much area as possible. It is normal for the exposure problem to exist over several floors with this type of fire.

Upper-story exposures can also be protected by working from within the exposed building. All exterior openings on the exposure side should be closed and, if needed, protective lines should be set up within the exposed building. It is best to lay several lines into the dry standpipe system of the exposed building whenever protective lines are to be set up from the interior of exposures over four stories in height. This will ensure an adequate supply of water at various floors for the use of the protective streams.

It is possible for an external exposure problem to exist above the reach of the streams from fire department equipment. When this situation exists, the protection of the exposures will be limited to the closing of external openings and the setting up of protective streams from the windows for the exposure. (See Figure 6.4.) This will require more streams than if the protective streams from the exposed building are used in conjunction with lines wetting down the exterior of the exposure. In such situations it should be remembered that closed windows do not fully protect the interior of the building from radiated heat.

There is one last point to discuss on the protection of external exposures. Hose streams set up for the protection of exposures do not normally need to be directed continuously on the exposure. Although the lines should be maintained in position to provide the necessary protection to the exposure, the stream can periodically be directed onto the main body of fire and redirected to the exposure as the need arises.

Preventing Interior Spread

Preventing the interior spread of the fire is primarily a matter of stopping its progress both horizontally and vertically. The problem is normally somewhat simpler in a one-story building than it is in a multiple-story building because it is only necessary to stop its horizontal progression. However, regardless of the size or construction of the building, each fire is an entity within itself, and no two are exactly alike. Consequently, it is not practical to try and establish hard and fast rules for the placement of hose lines used to cut off the progress of a fire; however, it is possible to examine some techniques that have proven to be valuable in confining an interior fire.

To stop the internal spread of the fire it is necessary to get cut-off lines into position at every location where the fire will travel if it is not stopped or extinguished. As a review, fire will travel horizontally in all directions and will travel vertically up any unprotected opening. However, this does not mean that it is necessary to bring lines in from all sides at every fire, as the fire is normally prevented from spreading in one or more directions by walls, fire doors, and so on. For example, in Figure 6.5 the fire will will spread horizontally only to the east, whereas in Figure 6.6 it will spread both southerly and easterly. In the situation shown in Figure 6.5 it will be necessary to bring lines in from the east to stop the progression, but in Figure 6.6 it will be necessary to bring lines from both the east and the south.

When making plans to get lines into position to stop the spread, it is important

Figure 6.4 Spray streams provide effective exposure protection

to remember that it takes time to lay lines and move them into position. Consequently, it is essential to estimate how far the fire will travel while lines are being laid and moved into position. If the estimation is faulty, the fire will probably progress past the point where it was planned to stop it. If this happens, additional lines will probably have to be laid and in some cases additional companies called to the fire.

If fire has an opportunity to travel both horizontally and vertically, its primary movement will be in a vertical direction. Not only will it move first in a vertical direction but it will also travel more rapidly in this direction due to its natural tendency to rise. Consequently, it is good practice to put the first hose line on the floor above the fire to cut off the vertical spread. Standpipes should be used for this purpose if they are available, as this makes it possible to provide lines to other floors

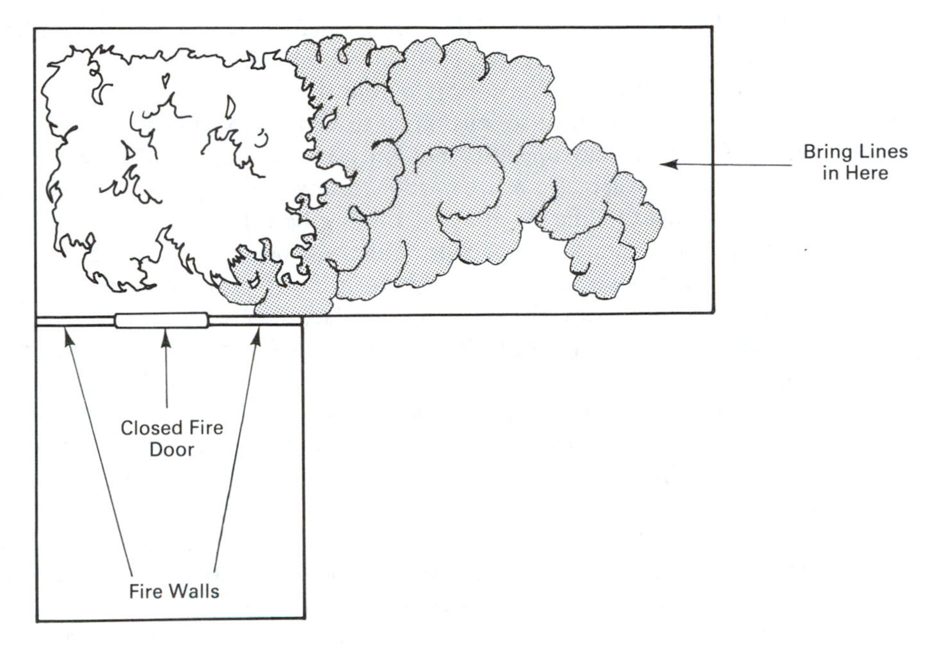

Figure 6.5 Walls and fire doors prevent horizontal fire spread

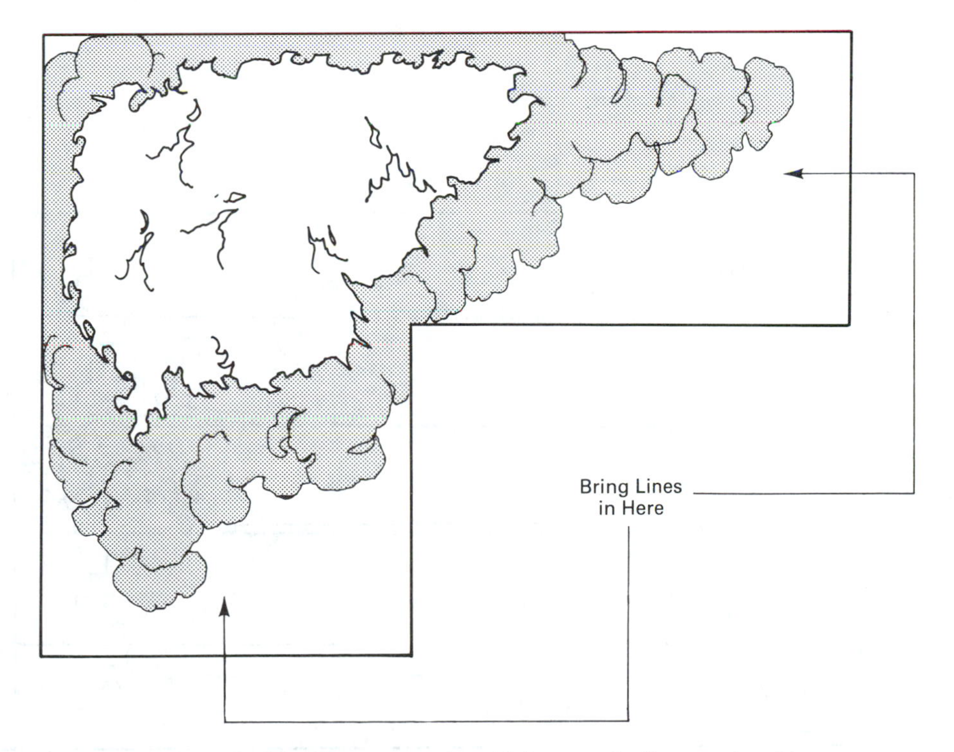

Figure 6.6 Lack of building construction aids increase the fire spread problem

with little additional effort if they are needed. Extreme care should be taken when advancing lines above the fire. Firefighters should not be placed in a position where their retreat path could be cut off by the fire.

Normally the most serious threat to vertical extension of the fire is open stairways. These vertical openings not only present an immediate danger from fire spread but they are also a life hazard to anyone caught above the fire floor. It is important that the line taken above the fire be used to gain control of one of these openings. Sometimes two or more lines will be required for complete control, in case more than one stairway extends from floor to floor. It is generally best to use a spray stream to stop the fire extension (see Figure 6.7); however, the line should not be opened until the heat developed on the lower floor becomes a threat to the upper floor.

If the fire is spreading simultaneously in two or more directions, it will generally be necessary to make a decision as to where to place the first horizontal cut-off line. In making this decision, it is necessary to consider the speed of the fire spread, how far it can spread, the potential life hazard in its path, and what type of material will eventually become involved if the fire is not stopped. A general rule is that the first cut-off lines should be positioned to protect life. Second priority should be given to the protection of any hazardous materials in the path of the fire. Not only will the control problem be intensified if these materials become involved but their burning may present a serious threat to the health or lives of the firefighters.

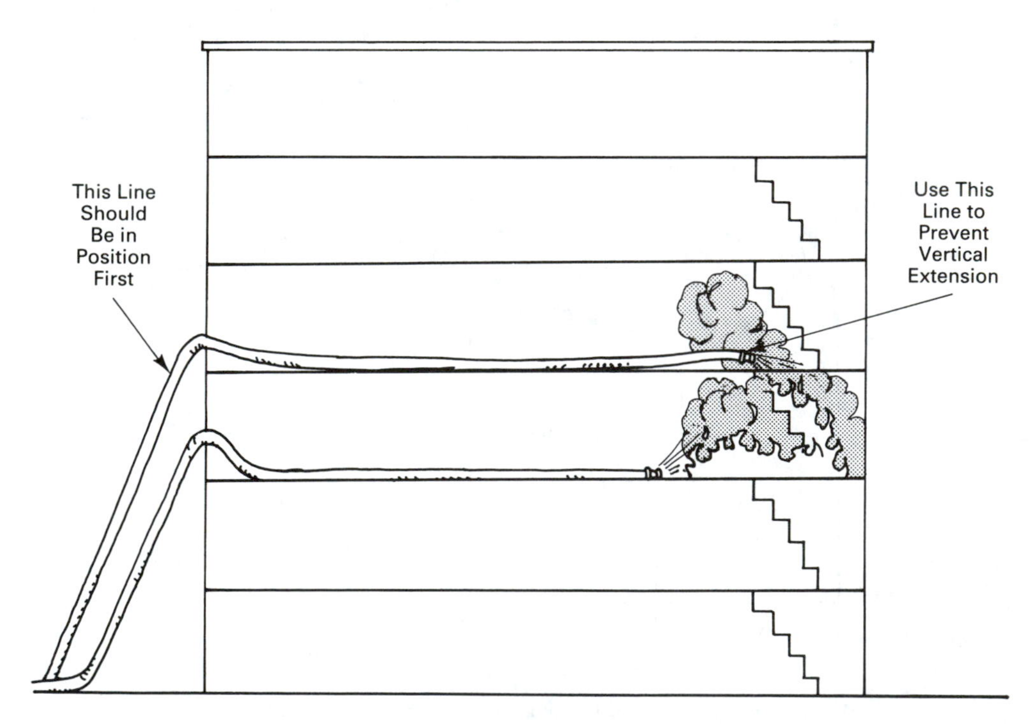

Figure 6.7 Gain control of the vertical openings

If there is no life hazard and no hazardous materials in the path of the fire, then it is good practice to place the first cut-off line in position to save the greatest portion of the building. For example, in Figure 6.6 the first line should be positioned to protect the eastern portion of the building. There is, however, an exception to this concept. If there is extremely valuable contents in the smaller portion of a building, then the first cut-off line should be positioned to protect this area. As an example, if the fire loss would be much greater if the southern portion of the building in Figure 6.6 was destroyed than if the eastern portion was lost, then the first cut-off line should be positioned to protect the southern portion of the building.

EXTINGUISHING THE FIRE

Most fires encountered by fire departments are of the Class A type. The primary extinguishing agent for Class A fires is water. Tactics used for extinguishment when water is used basically consist of determining the type of attack to be made, selecting the number and size of hose lines that will be required to extinguish the fire, and moving the lines into positions that will extinguish the fire most rapidly with the least amount of damage.

Water Application

During firefighting operations, water is generally projected onto the fire by the use of fire streams. A *fire stream* may be defined as a stream of water from the time it leaves a nozzle until it reaches the point of intended use, or until it reaches the limit of its projection, whichever occurs first. As it leaves the nozzle, the stream is affected by such factors as nozzle pressure, nozzle design, and, with adjustable nozzles, the adjustment setting. After it leaves the nozzle it is affected by such factors as wind, gravity, and air currents.

There are a number of different types of nozzles available for stream projection; however, all of them provide two basic types of streams: straight streams and spray (fog) streams. *Straight streams* may be produced by both smooth bore tips or by spray nozzles adjusted for straight stream application. *Spray streams* are produced by spray nozzles and fog nozzles.

This selection of the size and number of lines to be used for the attack is a matter of determining the amount of water that will be required to remove heat from the fire faster than it is being generated.

The amount of water discharged from a nozzle tip depends on the size of the tip and the discharge pressure. It is impossible on the fireground to determine the exact amount of water being projected from various streams at any one moment; however, there are some guidelines that can be used to achieve the desired results.

The discharge from 1-inch lines is approximately 30 gallons per minute (GPM). These lines are effective for extinguishing fires in trash and grass fires.

The discharge from 1½-inch lines is approximately 125 GPM whereas the discharge from 1¾-inch lines is about 175 GPM. Using two 1½- or 1¾-inch lines is normally sufficient to knock down a fire in one or two rooms of a dwelling or fires of similar size in commercial or industrial buildings.

A standard hose stream from a 2½-inch line will produce about 250 GPM. A 2½-inch line is the largest handline used in the fire service. These lines should be applied to large fires in commercial and industrial occupancies and, in some cases, on dwellings that are well involved. They are particularly effective on well-involved dwellings when spray nozzles are employed.

Some departments prefer to use 2-inch lines rather than 2½-inch lines. The 2-inch lines produce large volumes of water when equipped with automatic nozzles and are more mobile than 2½-inch lines.

Standard tips used on heavy stream appliances discharge a minimum of 500 GPM. Both straight and spray tips are available for use on heavy stream appliances. A 1⅜-inch straight tip will discharge approximately 500 GPM, a 1½-inch about 600 GPM, a 1¾-inch about 800 GPM, and a 2-inch about 1,000 GPM. Discharge from spray tips vary depending on their size, setting, and nozzle pressure. Some spray nozzles produce a constant volume of water.

Good judgment must be used when selecting the size of line to be used. All the facts must be considered when weighing the value of a small line against that of a large one. Small lines offer the advantage of mobility and ease of operation, whereas large lines offer volume, reach, and striking power at the cost of reduced mobility. One of the basic principles that should be considered when making the decision as to the size of line to use is *never underestimate the size of the fire.* It is

Photo 6.3 Each of the heavy streams shown here is throwing a minimum of 500 GPM into the fire area. (Courtesy of Chris Mickal, New Orleans Fire Department Photo Unit)

better to overkill than have to back out because there is not sufficient water to do the job.

Attacking the Fire

There are two basic types of attacks that can be made on the fire for the purpose of extinguishment: direct attack and indirect attack.

The direct attack. A direct attack is one where water is applied directly to the fire. The theory behind the direct attack is that the best method of extinguishing the fire is to eliminate it at its source, or what is referred to as the seat of the fire. The seat of the fire, as defined earlier, is the hottest part of the fire and is usually found at the point of origin. Since a large amount of heat is being generated at this point, water applied at this location will exert its most effective cooling action.

Both straight streams and spray streams are used to make a direct attack on the fire. Spray streams are normally used when 1½- or 1¾-inch hand-held lines are used in a direct attack. A 1½-inch line can physically be handled by one firefighter; however, for safety purposes it is best that two firefighters be used to man both 1½- and 1¾-inch lines. The spray stream provides better protection for the firefighter and permits a closer attack on the seat of the fire than when straight streams are employed.

The direct attack method is almost universally used on large fires that have gained considerable headway, such as lumberyard fires, fires that have burned through the roof, fires in structures that have burned through to the outside, or fires that are well involved. With these types of fires, both straight streams and spray streams are used on hand-held 2- and 2½-inch lines. Both types of streams are also used when heavy stream appliances are placed in operation. To be able to effectively maneuver a 2½-inch hand-held line requires a minimum of three firefighters.

The straight stream offers distinct advantages in this type of attack. It has the reach, penetration, and striking power necessary to bore into the fire through radiant heat that would boil off a spray stream before it reached the seat of the fire. A straight stream, however, forces the firefighter to operate from some distance back due to the heat produced by the fire. Large amounts of water are required on fires of this type since the water strikes the burning material and runs off too quickly to absorb a maximum quantity of heat. As a result, water damage may be considerable.

Many times a direct attack on the fire is made by the use of heavy stream appliances. A heavy stream appliance is a piece of equipment that discharges more water than can be projected from a hand-held line. The minimum amount of water normally discharged from a heavy stream appliance is about 500 GPM; however, nozzles producing a lesser amount of water are sometimes substituted for the standard tips. Streams from these appliances are referred to both as *heavy streams* and *master streams.* Heavy streams take time to set up, but there should be little hesitation to use them if the volume of the fire warrants or if life hazard is a factor. They are excellent for making a quick knockdown of the main body of fire, thus providing a means for gaining access to the fire by the use of handlines. It is good practice to use heavy streams only as long as necessary, and *never* after firefighters with handlines have entered the area.

One of the basic tactical principles involved with the direct attack is to bring in hose lines from the uninvolved portion of the building and push the fire back into the involved portion. Although this principle seems simple, it is one of the most often violated. The result is that fire is frequently pushed unnecessarily into uninvolved areas of the building. The reason the principle is violated is that firefighters are basically aggressive and want to hit the fire where they see it. For example, a company pulls up in front of a dwelling that has one of the front rooms well involved with fire. Fire is shooting out the windows in front and on the side. Firefighters lay lines and fight their way to one of the windows, hitting the fire as they proceed. When they reach the window they direct the stream inside in an attempt to achieve a fast knockdown. The result is that they probably pushed the fire and its smoky by-products to other portions of the building. The proper tactic is illustrated in Figure 6.8.

One line is taken in through the front door and advanced down the hallway to the fire area. The fire is aggressively attacked and pushed out the window while being extinguished. The outside line is protecting the exposures, which includes non-burning portions of the involved building.

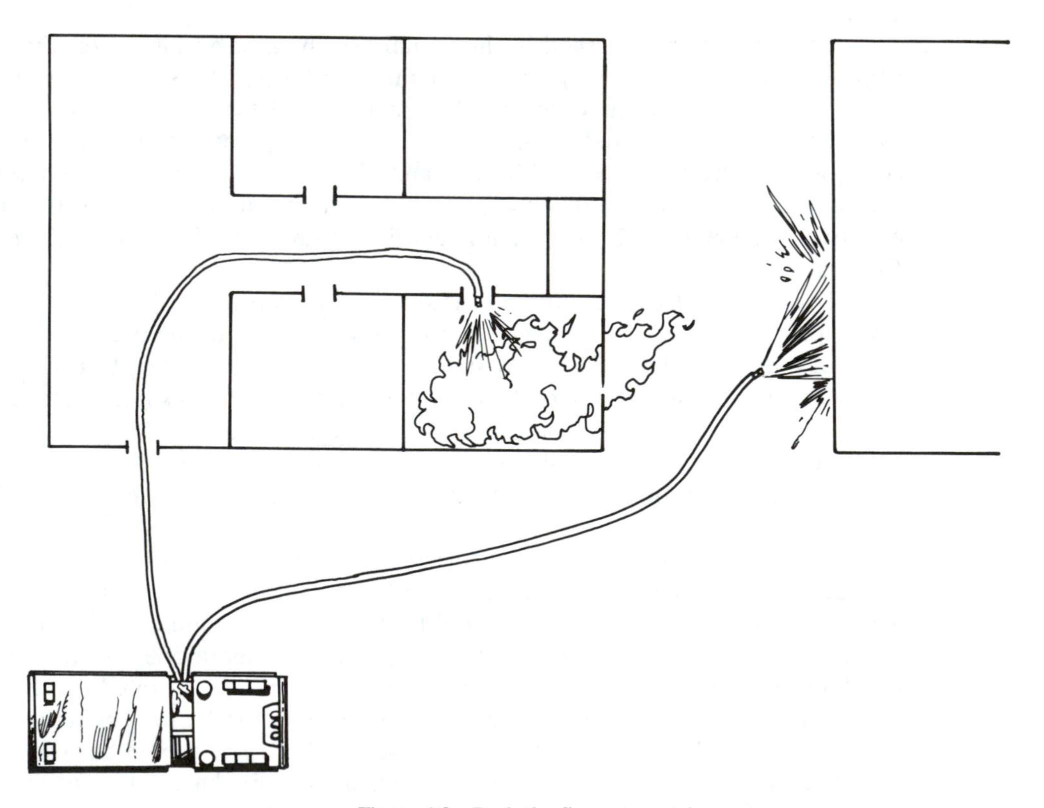

Figure 6.8 Push the fire out—not in

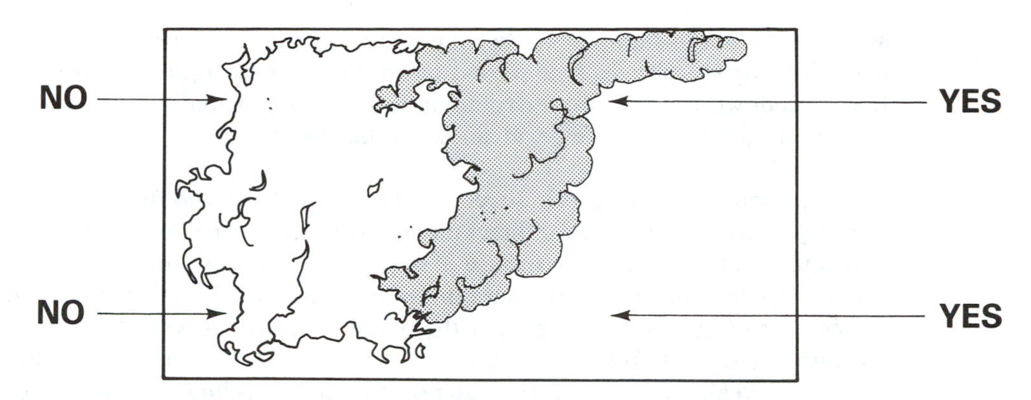

Figure 6.9 Bring lines in from the uninvolved portion of the building

Another example of this tactical principle is illustrated in Figure 6.9. One end of an industrial building is well involved with fire. Lines are brought in from the opposite end and the fire is attacked. This procedure restricts the damage to the involved portion of the building. If lines had been brought in from the opposite end, the fire would probably have been pushed throughout the entire area and the building lost.

The principle should also be kept in mind for fires in multiple-story buildings. Figure 6.10 illustrates a room fire on the third floor of a five-story hotel. Two lines

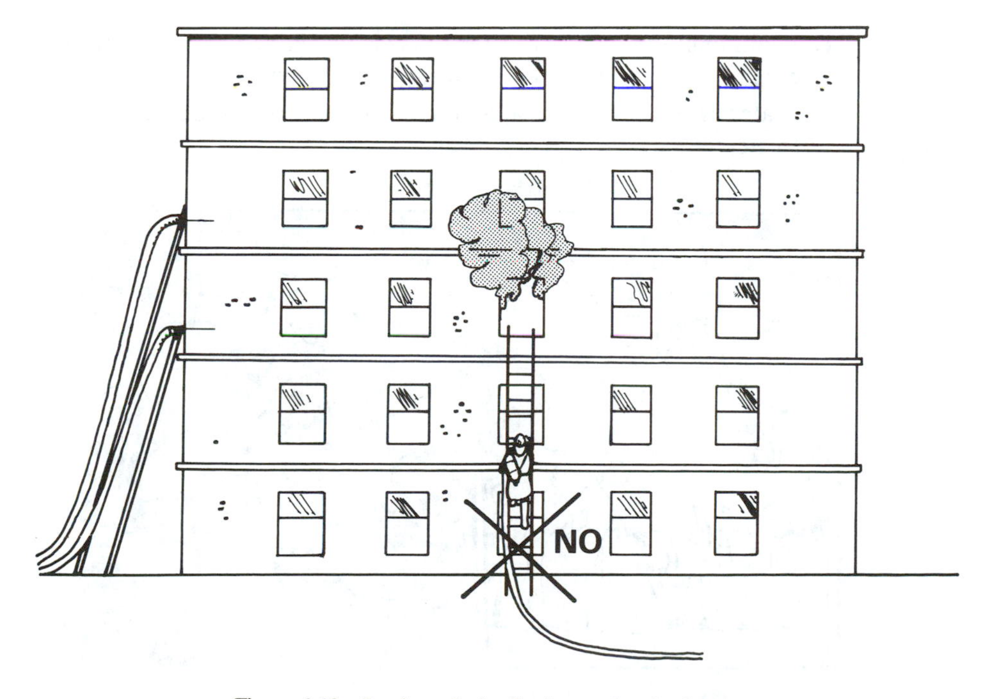

Figure 6.10 Don't push the fire into uninvolved areas

are being advanced simultaneously. One is taken from the fourth floor to gain control of the open stairway. The second has been advanced up a ladder to the third floor and down the hall into the room on fire. A ladder should not be raised to the fire room nor should a line be advanced in this direction.

The indirect attack. Indirect attacks are made on the fire through the use of spray or fog streams. The indirect method is based on the theory of securing the maximum cooling effect from the use of the water applied. With the indirect attack, water is applied under high pressure in the form of a wide-angle spray or in a fog pattern. The water is not directed at the seat of the fire, as with the direct attack, but rather above the fire and into the heat that has built up at the ceiling level. The line is swirled about to project the water over as great a ceiling area as possible. The object is to cool the fire area to a temperature below the ignition temperature of the material burning. (See Figure 6.11.)

With the indirect method of attack, theoretically each drop of water soaks up the maximum amount of heat and is converted into steam. The extinguishing effect is somewhat twofold:

1. The temperature is lowered so rapidly that heat is absorbed faster than it is being generated.
2. Some degree of smothering effect is produced by the steam.

Some of the advantages claimed for the indirect method of attack are:

1. A minimum amount of water is used for extinguishment since the maximum cooling is being obtained from the water used.
2. Water damage is held to a minimum since nearly all the water used is converted to steam.

Figure 6.11 The indirect method of attack

3. Firefighters are subjected to far less punishment since the attack is normally made from outside the fire area, such as from a doorway or window.
4. A faster knockdown is generally achieved.

Two conditions are essential for the most effective application of the indirect method. First, it must be a fire that is generating a considerable amount of heat. Second, the fire must be burning in an area that is not vertically vented and preferably one that is not vented at all.

Selecting the proper method of attack. The choice of the type of fire stream, the size of the stream, and whether a direct or indirect attack is to be made on the fire will depend on the total knowledge gained during the size-up. The question that must be answered is: Which configuration will extinguish this particular fire with the least amount of loss to life and property? Basically, the number and size of lines required and whether a direct or indirect attack will be made will be determined by the amount of heat present, where the fire is, and where it is likely to go.

On interior fires, if the heat will not prohibit entering the building, small lines of 1½ or 1¾ inches will usually suffice. These lines discharge a sufficient volume of water to handle the fire, and will permit the degree of maneuverability desired. However, larger lines should be placed into operation without hesitation if there is any doubt as to the ability of small lines to control the fire. It is easier and quicker to reduce a line than it is to lay a new one. When selecting the size of lines to use, consideration should always be given to the progress the fire will make during the time required for placing the hose lay into operation.

REVIEW QUESTIONS

1. What is the difference between firefighting strategy and firefighting tactics?
2. What type of future changes should be expected in firefighting strategy and tactics?
3. In its simplest sense, what is basic firefighting strategy?
4. Tactically, what does locating the fire include?
5. What is meant by the seat of the fire?
6. Where is the seat of the fire normally located?
7. How much truth is there in the saying, "Where there is smoke there is fire"?
8. How effective is the sense of smell in helping to locate the source of smoke?
9. What is live smoke?
10. What is dead smoke?
11. How should an electric motor be checked to see if it is the source of smoke?
12. If the entire upper floor of a multiple-story building is filled with smoke, what other area should be checked as a potential location of the fire?
13. When searching for the source of smoke, how should an answer from an occupant such as "There's no fire in here" be handled?
14. What types of materials should be expected to be involved in fires in dwellings, office buildings, and small mercantile establishments?

15. What might it indicate if a fire hotter than normal is encountered or the fire moves faster than expected?
16. What color of smoke is usually produced by materials such as roofing paper, asphalt, rubber, and petroleum products?
17. What would abnormal colors in the smoke such as yellow and red generally indicate?
18. What color of smoke is generally produced from a fire in a normal room containing furnishings and painted walls?
19. What color of smoke is generally given off from a house or garage fire?
20. What is the best clue as to where the fire will spread?
21. Of the four methods by which heat travels from one body to another, which three play the biggest part in the rapid travel of a fire?
22. Of the four methods by which heat travels from one body to another, which one has the most influence on the vertical travel of the fire?
23. What is the basic principle involved when making plans as to where to make a stand to stop the spread of the fire?
24. From a theoretical standpoint, what is meant by confining the fire?
25. From a practical standpoint, what are the two objectives that should be established for the tactics to use to confine a fire in a structure?
26. How is the objective of preventing the spread of the fire from the building of origin, or from the buildings involved at the time of arrival of the fire department generally achieved?
27. How is the task of keeping the temperature of the exposure below its ignition temperature generally achieved?
28. How are the two objectives for confining the fire generally achieved at most fires?
29. At all fires, what is one of the most effective methods of protecting the exposures?
30. What is important at a well-involved dwelling fire where both of the initial lines are used to protect the exposures?
31. What is the big difference between the tactics used to protect the external exposures at a fire in a dwelling and one in an industrial building?
32. What method is used to compensate for the lack of operational experience of most firefighters at large fires?
33. Where is it best to direct the stream when wetting down an exposure?
34. What is the difference in the ability of an engine company to protect the exposures at a well-involved dwelling fire and a large fire in an industrial plant?
35. What methods are used to protect the exposures from a large fire in the upper floors of a multiple-story building?
36. What is a water curtain?
37. What is the effectiveness of a water curtain as compared with the wetting down of an exposure?
38. What other method is available for protecting exposures from a fire in a multiple-story building other than a water curtain and wetting down the exposure?
39. What is necessary in order to stop the internal spread of a fire?
40. What does the fire travel have to do with the making of plans to stop the internal spread?

41. What is considered good practice regarding the first line to be laid to protect the internal spread in multiple-story fires?
42. Normally, what is the most serious threat to vertical extension of the fire?
43. If two lines are necessary to prevent the horizontal spread of the fire, what has to be taken into consideration when determining where the first line should be taken?
44. If two lines are needed to prevent the horizontal spread of the fire, what factors should be given first and second priority?
45. What is the definition of a fire stream?
46. What are the two basic types of fire streams?
47. What factors should be considered when selecting the size and number of lines to be used for an attack on the fire?
48. What two factors determine the amount of water discharged from a nozzle tip?
49. Approximately how much water is discharged from a 1-inch line? A $1\frac{1}{2}$-inch line? A $1\frac{3}{4}$-inch line? A $2\frac{1}{2}$-inch line?
50. What is the minimum amount of water discharged from a standard tip used on a heavy stream appliance?
51. Approximately how much water is discharged from the following straight tips: $1\frac{3}{8}$ inches? $1\frac{1}{2}$ inches? $1\frac{3}{4}$ inches? 2 inches?
52. What is the basic principle that should be considered when making a decision as to the size of line to use on a fire?
53. What are the two basic types of fire attacks?
54. Where is the water directed on a direct attack?
55. What is the theory behind the direct attack?
56. What type streams are used on a direct attack?
57. On what types of fires are direct attacks almost universally used?
58. What is the minimum number of people required to be able to maneuver a $2\frac{1}{2}$-inch hand-held line effectively?
59. What is considered as good practice in regard to the length of time that heavy stream appliances should be used at a fire?
60. Where should lines be brought into the fire building when a direct attack is being used?
61. Where is the water directed when making an indirect attack on the fire?
62. What is the theory behind the indirect attack?
63. What is the objective of the indirect attack?
64. What is the twofold extinguishing effect of the indirect attack?
65. What are the two conditions essential for the most effective application of the indirect method of attack?
66. What question should be answered when determining whether to use a direct or indirect attack on a fire?

CHAPTER 7

Fire Command

OBJECTIVE

The objective of this chapter is to examine the various aspects of fire command. The actions required by a company commander prior to the alarm, while responding, and upon reaching the fireground are explored. A brief examination is made of the incident command system as it develops from a first alarm to a multiple alarm situation. Review questions are provided at the end of the chapter to assist the reader in determining his or her level of comprehension of the chapter contents.

It is important that a firefighter have a basic understanding of the thought process the company commander goes through as that individual makes a decision as to the action to take at a fire. Knowledge of the thought process provides the firefighter with a better understanding of the reason that certain decisions are made. This understanding makes it more palpable for a firefighter to carry out an order that otherwise might seem questionable.

BEFORE THE ALARM SOUNDS

Theoretically, some of the information upon which a company commander makes a decision at a fire is gathered when the company makes a pre-fire planning inspection of the building in which the fire occurs. Unfortunately, although this is the ideal, a pre-fire planning inspection has not been made in the buildings where a great majority of the fires occur. Pre-fire planning inspections are normally limited to a company's target hazards. A *target hazard* might be defined as an occupancy that constitutes a large collection of burnable values or an occupancy where the life hazard is severe. These are the occupancies where the maximum fire prevention effort of a department is concentrated, and the occupancies where the strictest laws are generally applicable regarding built-in fire protection. As a result of the combination of strict laws and maximum enforcement, fires are kept to a minimum. A large majority of all fires that occur in most cities are in buildings other than target hazards.

WHEN THE ALARM SOUNDS

The first information a company commander receives regarding a fire in a building is generally extremely limited. An alarm sounds throughout the engine house and a message similar to the following is received:

> Engine 5, Engine 7, Truck 8, and Battalion 3—Respond to a structure fire at 347 W. Maple.

If the dispatcher has received additional pertinent information such as "The fire is reported on the third floor," or "It is reported that people are trapped in the building," he or she might pass it on to the companies.

Prior to leaving quarters, the company commander of an engine company will generally check the map to review the location of the last hydrant available before arriving at the given location, and the next available hydrant past the given location. (See Figure 7.1.) The company commander will also check to see if there is an alley to the rear of the address and also behind the addresses across the street. The commander knows by the address received whether the given location is on the north,

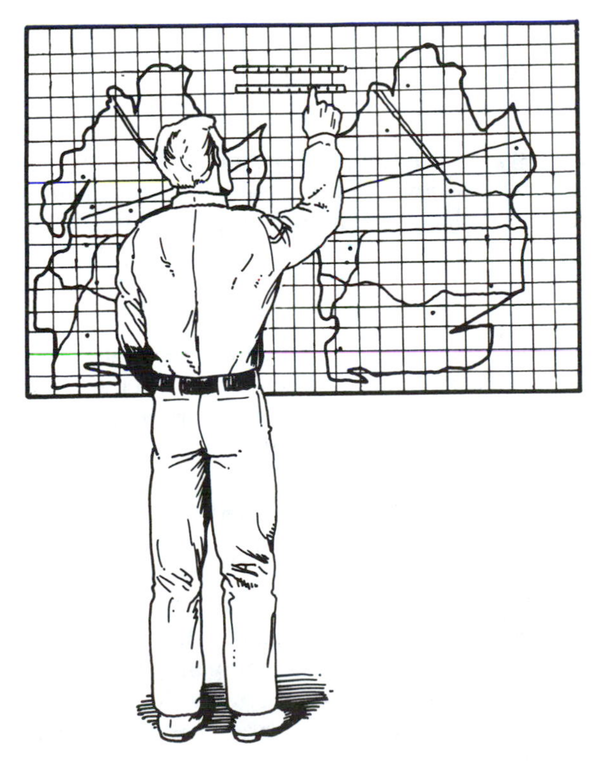

Figure 7.1 Check the map before leaving quarters

south, east, or west side of the street; however, the commander also knows that many times a neighbor across the street will call in the alarm and give his or her own address.

RESPONDING TO THE ALARM

There are a number of thoughts that go through the company commander's mind as the apparatus pulls out of quarters. By mentally reviewing the type of area to which the department is responding, the commander has a general idea of whether the fire is in a single-family dwelling, a commercial occupancy, or whatever. The mental review also takes into consideration what affect the time of response and the weather will have on the life hazard, the traffic problems, and the fire spread. If the address is in the company's first-in district, the commander knows that his or her company will probably be the first to arrive and he or she will be in charge until the chief arrives. The commander's thoughts consider what other companies will be responding and from what direction they will be approaching the fire location. He or she mentally calculates what companies will arrive on a second alarm, if the fire is of such a size to warrant one.

While responding, the company commander's eyes are constantly searching in the direction of the given location—watching for any sign that they may have a "worker" (a fire in progress). If smoke is sighted, the size and the color will give some clue as to what is burning.

THREE POSSIBILITIES

One of three conditions (or a slight variation) will normally be observed by the company commander as the apparatus approaches the last available hydrant. The commander will have to make an initial decision prior to passing that hydrant.

Condition Number One

One of the three conditions is that nothing will be showing. (See Figure 7.2.) Under this condition the company commander should proceed directly to the reported address. While moving from the hydrant to the address, a report should be made to the dispatch center. The call might go somewhat as follows:

> Dispatch from Engine 5. On the scene at 347 W. Maple. Nothing showing. Will investigate.

This report will alert the dispatch center that a company is on the scene and will provide additional information to incoming companies. Upon hearing this report, the company commander of the second engine company should proceed to the last hydrant available and wait for further instructions.

There normally will be someone to meet the apparatus as the company pulls

Figure 7.2 Condition number 1—Nothing showing

up to the site. The company commander will listen to what is said and then go inside to check the reported condition. It is good standard practice for a firefighter to take a water extinguisher and proceed inside with the company commander. In the majority of cases where nothing is showing on arrival, if there is a fire it can be extinguished by the use of the hand extinguisher.

It is possible that the company commander may go inside and find a small fire that will require the use of a 1½- or an 1¾-inch line for extinguishment. In this case the commander will return to the apparatus and call for the needed line. He or she is in charge and will take command of the situation. If the commander feels that a backup line is needed he or she will instruct the company waiting at the hydrant to lay one. It is good practice for the first-in officer always to have a backup line laid whenever a working line of any size is being used in the building. If the fire is of sufficient size, the commander may decide to have a company lay a line to the rear of the dwelling. In this case the commander would instruct the company officer waiting at the hydrant or perhaps the third-in engine company commander to do so. Existing conditions should be reported to the dispatch center as soon as practical.

Condition Number Two

Another situation that the company commander may observe upon approaching the reported address is a small amount of smoke or fire (or both) in the direction where the fire was reported. (See Figure 7.3.) Prior to passing the hydrant the commander will have to make a decision as to whether to lay a large line or proceed to the fire and take it with the tank. There are no hard and fast rules on which the individual can base a decision. The decision will be left to one's subjective judgment, using the basis of keeping the fire loss to a minimum. If the commander decides to handle the fire with the water in the tank, a backup line should always be laid. Regardless of the decision, the commander should report the condition to the dispatch center. The report may go something like this:

Figure 7.3 Condition number 2—Smoke showing

> Dispatch from Engine 5. At 347 W. Maple, we have a small fire in a single-family dwelling. We will hold the first alarm assignment. Engine 7, lay me a backup line.

Incidentally, if the fire happens to be at an address different than originally received, the correct address should be reported to the dispatch center.

In this situation, as with all situations, the company commander of Engine 5 should remember that he or she is in charge of the fire until relieved by a superior officer. Not only would it be important to instruct Engine 7 to lay a backup line but he or she should also instruct the company commanders of the other arriving companies as to where they should position themselves.

Condition Number Three

The third condition that the company commander may encounter is one where there are definite signs of a fire as the reported address is approached. (See Figure 7.4.) The commander will then have to decide if the fire can be controlled by the first alarm assignment or if help will be needed. In either case the commander will lay a line going in and proceed to attack the fire. If the individual decides that the fire can be controlled by the first assignment, the report to the dispatch center will go something like this:

> Dispatch from Engine 5. We have one room of a one-story, single-family dwelling well involved. We will hold the first alarm assignment. Engine 7, lay a line down the alley and protect the exposures to the rear.

Being in command, the commander should also instruct the other incoming company commanders as to what they should do.

Figure 7.4 Condition number 3—Fire showing

MULTIPLE-ALARM FIRES

It may be apparent some distance from the reported location that the fire is of considerable size and probably beyond the control of the first alarm assignment. In such situations there is no harm to be done by ordering a second alarm assignment. The actual call for the number of companies desired, however, should be reserved until an adequate on scene evaluation can be made.

Taking Command

In most cases the company commander of the first-in company will arrive on the scene prior to additional help being requested. The company commander, being the first to arrive, is in charge until relieved by a chief officer. At a large fire, the commander should carefully evaluate the situation, turn his or her company over to the acting officer, and take command of the fire. It is poor practice to wait for the chief to arrive to make the necessary decisions.

One of the first acts should be to request the additional help that the company commander deems necessary, keeping in mind the progress the fire will probably make before additional companies arrive on the scene. It is not good practice merely to call for a second or third alarm. When this is done, the company commander is basically saying, "I know that help is needed, but I'm not really sure how much." The company commander should also keep in mind that there is every excuse for overestimating the amount of help needed, but little excuse for underestimating. It is much more professional to return companies to quarters because they are not needed than to later have to call for additional help because an insufficient amount was not originally requested.

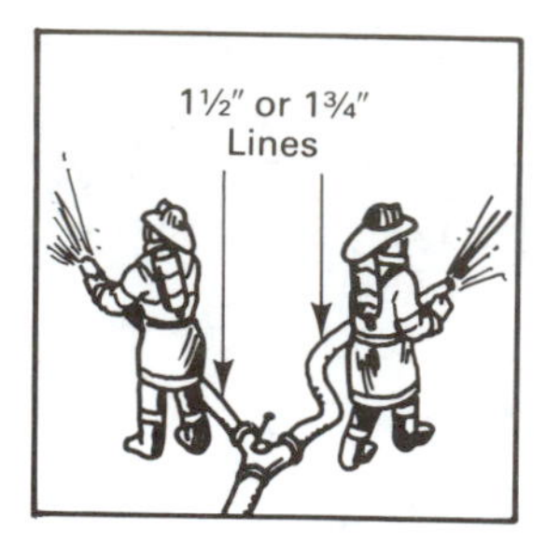

Figure 7.5 The capabilities of an engine company

Estimating the Number of Engine Companies Needed

All officers should have some guidelines that they use to estimate the number of companies required to control the fire adequately. There is no single system that can be considered as best. Different officers use different systems with similar results. The important thing is to have a system that is practical and workable. The following might be considered in the development of a system. (See Figure 7.5.)

Most engine companies are capable of one of the following:

1. Manning two 1½- or 1¾-inch lines.
2. Manning a single 2- or 2½-inch line.
3. Putting a heavy stream appliance into operation.

Photo 7.1 A 1½-inch line can effectively be handled by one firefighter.

Using this information as a guideline, an officer can quickly estimate the number of engine companies required at the fire. For example, if an officer estimates that two $1\frac{1}{2}$- or $1\frac{3}{4}$-inch lines and two $2\frac{1}{2}$-inch lines will be needed for extinguishment, and that two $2\frac{1}{2}$-inch lines and a ladder pipe will be needed to protect the exposures, the officer knows that he or she will need six engine companies for fire control. It is good practice as a precautionary measure to order one more company than is estimated for control. It is also good practice at a fire where wood shingle roofs are involved to have an engine company patrolling downwind from the fire as a precautionary measure against flying brands. Consequently, in the above example, a total of eight engine companies would be required—six at the fire, one patrolling downwind, and one as a precautionary measure. If the officer receives three engine companies on the first alarm assignment, the situational report to the dispatch center would include a request for five additional engine companies.

It might be noted the above guidelines are normally not satisfactory for use at high-rise fires, basement fires, or other fires that will extend for a long period of time. These fires normally require that the firefighters at the end of the line be relieved periodically because of the limited time available on their breathing apparatus. These fires generally require three engine company crews for every manned line—one working on the fire, one waiting to relieve those working on the fire, and one changing bottles on the breathing apparatus. For example, in the previous illustration where six engine companies were needed at the fire, if the two $2\frac{1}{2}$-inch manned lines fell into the rotation category (3 for 1), then an additional four engine companies would be needed at the fire. This would bring the total to twelve engine companies—ten at the fire, one patrolling downwind, and one as a precautionary

Photo 7.2 Engine company members man both portable heavy stream appliances and handlines. (Courtesy of Chicago Fire Department)

measure. The initial request would then be for nine additional engine companies. (See Figure 7.6.)

Estimating the Number of Other Companies Needed

Consideration should also be given as to the number of truck, salvage, and rescue companies needed at the fire. It is generally more difficult to estimate the number of these companies required than it is to estimate the required number of engine companies. Regardless, all officers should have some guidelines they can use at a fire to determine the need for these companies rather than haphazardly pulling a number out of the air. The following is offered as a guideline for assistance in the development of a workable system.

Figure 7.6 The 3-for-1 situation

At structure fires, one truck company is normally needed for every three engine companies at the fire in order to carry out the duties assigned to the truck company. (See Figure 7.7.) If the life hazard is extreme, one truck company for every two engine companies may be required, and in some cases a one-for-one match would not be excessive. There should be a minimum of one truck company at every structure fire of any size. If a sufficient number of truck companies is not available, it is good practice to request additional engine companies and assign them the duties that are normally the responsibility of the truck company.

At least one salvage company or a company doing salvage work is needed whenever lines are being used above the first floor. One company doing salvage work is necessary for approximately every three engine companies working inside the building. Extra engine companies should be requested to perform salvage operations in those cities not having separate salvage companies. Additional companies for salvage operations will probably be required if the use of water is excessive.

Emergency medical care capabilities should be available at every working fire. A company should be on the scene that has no other responsibilities other than that of performing emergency medical care. Ambulances should be ordered if there is any possibility they may be needed.

The Problems of Taking Command

The practice of the first arriving officer taking command of the fire is the exception rather than the norm in many cities. This is unfortunate, since the action taken during the first five minutes will have a major impact on the success or failure of

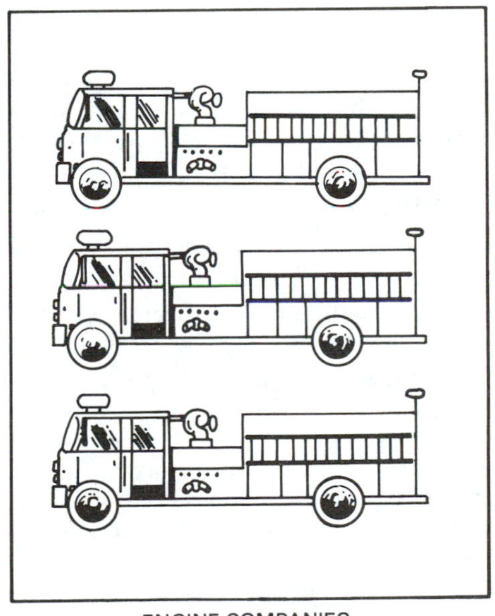

ENGINE COMPANIES

TRUCK COMPANY

Figure 7.7 The 3-to-1 ratio

firefighting operations. Although failure of the first arriving officer to take command normally does little harm at the majority of fires, its impact is felt at larger fires. Fire operations require a coordinated effort. Coordination is required whenever two or more units are working on a fire. Unless the first arriving officer takes command, all officers of first arriving units will place their company where they feel it can do the most good. In most cases their decisions are satisfactory, but not always. Many chief officers have been frustrated when arriving at a fire to find that the companies are not where they would like them to be. In fact, on many occasions it has been necessary for a chief officer to request additional companies to lay lines to needed locations because all of the first arriving companies were committed and had lines charged. A request for additional companies had to be made, despite the fact that there were a sufficient number of companies and personnel on the first alarm assignment to control and extinguish the fire adequately—providing that companies were committed to the correct locations.

Taking command of a large fire is an awesome responsibility. This is particularly true for a young company commander whose previous command experience may have been limited to directing his or her company at a single company fire. However, when an officer is given the responsibility for commanding, it is important that he or she take command. This means that the individual assumes the responsibility for the management of all operations, and has the authority to order as much equipment as deemed necessary. When assuming command, the officer takes on the full responsibility for planning, organizing, directing, coordinating, and controlling all the resources at the scene of the emergency.

Let's revert a step and consider that the first arriving company officer at a large fire evaluates the situation and recognizes that it cannot be controlled by the first alarm assignment. The individual recognizes and accepts the responsibilities of the position and takes command. The officer has turned his or her company over to the acting officer, sized-up the situation, and starts planning the strategy for extinguishing and controlling the fire. He or she has mentally evaluated the problems of rescue, laddering, ventilation, forcible entry, salvage, controlling the utilities, emergency medical care, and the laying of lines for extinguishment and protecting the exposures. The officer's evaluation indicated where companies will be needed to alleviate the problems, taking into consideration where the fire will probably be by the time the needed companies arrive. The evaluation further indicates that seven engine companies, three truck companies, two salvage companies, and one rescue company will be needed to control the fire effectively. The officer's first alarm assignment consists of two engine companies, a truck company, and a battalion chief. The report to the dispatch center may go somewhat as follows:

> Dispatch from Engine 3. At 937 South Main we have a two-story brick commercial building with the second floor well involved. There are serious exposures to the north and west. Give me five more engine companies, two more trucks, two salvage companies, and a rescue. I am establishing the command post in front of the building. Let me know what additional companies I will be receiving as soon as possible.

Note that the officer identified the location of the command post and requested information regarding which additional companies will be arriving. It is important that he or she write down this information as it is received, for it will assist him or her in taking the next command step—that of directing the incoming companies. This step should be taken without hesitation. Failure to do so would normally result in officers of the first alarm assignment placing their companies where they felt they were best needed. This may not be where the officer who took command wants them. This would require a complete reevaluation of the original strategic plans.

The directing and placement of companies does not have to be expressed in minute detail. The objective is to get the company where it is needed and have it carry out the duties desired. This is fire command strategy. How the duties are carried out should be left to the discretion of the company officer. This is fire company tactics. Following are a few examples of how orders may be given:

Engine 3, lay a line and protect the exposure to the south.

Truck 2, evacuate the fourth and fifth floors.

Engine 6, take the back of the building.

Engine 7, get a line above the fire.

Changing Commands

In most cases the district chief will have arrived on the scene by this time. It is important that the dispatch center be notified of the district chief's arrival so that the dispatch center and all officers at the fire know that there has been a shift in command.

It is now important that the company officer who took command brief the district chief on the entire strategic plan. The chief has undoubtedly picked up the radio message requesting additional companies and has mentally stored the information. After making his own size-up situation, the district chief may decide to go with the original plan or modify it. The company officer should not be offended if the latter is the chief's choice. The chief may have additional information regarding the occupancy or may have seen something that was not apparent to the company officer. The chief's years of experience may also have taught him a few tricks of which the company officer is not aware.

THE INCIDENT COMMAND SYSTEM

The entire procedure so far has described part of an overall plan known as the Incident Command System. The Incident Command System has been designed to provide for an organized method of achieving results on the fireground. It provides for constant command and coordination regardless of the size of the emergency.

There is a big jump in the Incident Command System as the fire progresses from one that can be handled by the first alarm assignment to that of a multiple-alarm assignment. All aspects of the first alarm fire are normally effectively handled by a district chief. The jump to a multiple-alarm fire requires the division of the fire into sectors and the addition of staff assistance to the incident commander. A multiple-alarm fire in most cities provides for the dispatch of a chief officer at the next higher level of command and at least one additional district chief. A good practice is to have a district chief on the scene for each five companies (or portion thereof) at the fire. Five companies is about the maximum that can be effectively supervised by one person in critical situations.

The nature of the command problem is apparent to the first responding district chief upon hearing the first arriving company officer's request for more companies. One of the first things that should be expected is that the media will be on the chief's back within a short period of time. Consequently, one of the first steps the chief will want to take after arriving on the scene is to appoint an information officer. This officer will be responsible for keeping the media informed, which will permit the incident commander to devote all thought processes to control and fire extinguishment. Generally there is no better person to appoint to this position than the first arriving officer who had originally planned the overall strategy and requested the additional companies. That individual is probably more aware of the overall problems than any other person on the scene. However, it is a good idea to replace this officer with a regularly assigned Public Information Officer or a staff officer as soon as possible and allow the first arriving officer to return to his or her company.

Once the district chief has been briefed on the first arriving officer's strategy, the chief should make and start implementing a plan for dividing the fire. (See Figure 7.8.) It is good practice, at a minimum, to have one person responsible for all activities inside the building and one for the activities to the rear of the building. If the fire is in a multiple-story building, it may be necessary to have a commander in charge of each floor. The districts chief's evaluation of how the fire should be divided may indicate a request for additional chief officers. However, prior to the

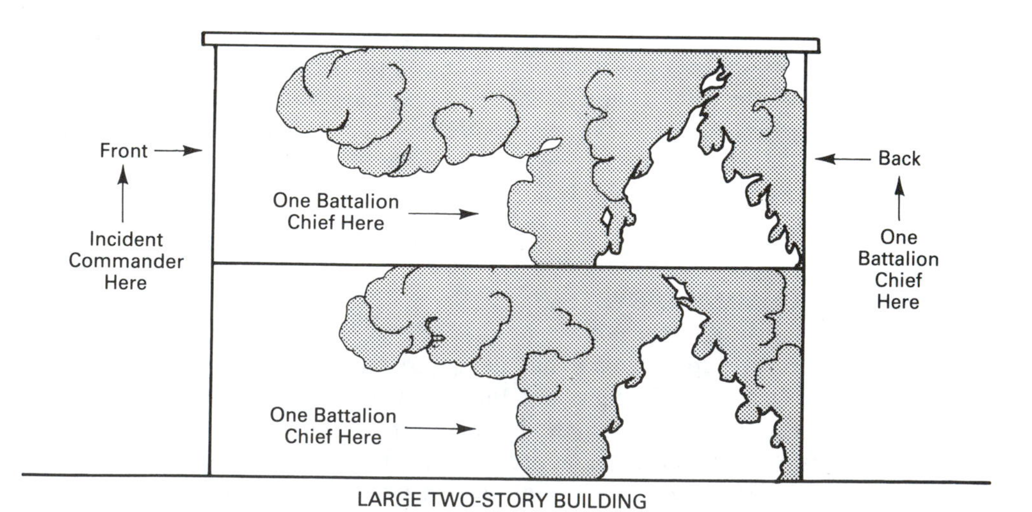

Figure 7.8 Divide the fire

arrival of additional chief officers, the district chief should assign company officers as section commanders.

Different cities have different titles for the next level of command. This chief officer may be referred to as a division chief or an assistant chief. Regardless of the title, the dispatch center should be notified of the division (assistant) chief's arrival on the scene and informed that that individual is taking command. This is important because ranking officers will many times appear at multiple-alarm fires and be there only for the purpose of observing. It should never be assumed that a ranking chief has taken command because he has arrived on the scene, and there should never be any doubt in anyone's mind as to who is in charge of the fire.

After the division chief has been briefed and has assumed the duties of incident commander, the district chief will be assigned to command a section of the fire, unless, of course, the incident commander decides to use the district chief in a staff position.

Large-Scale Emergencies

Fires that spread over a large area or extend over a long period of time complicate the task of the incident commander, who may now have to consider feeding personnel, refueling apparatus, adding special equipment such as helicopters, and coordinating activities with other agencies. If the incident commander attempted to cope with all these additional factors alone, he would be overwhelmed. The only logical alternative is to acquire a reliable staff and delegate some of those functional responsibilities to the individuals assigned. In doing so, the incident commander's span of control should be limited to a maximum of five people. (The actual number may be less, depending on the magnitude of the problem.)

The Incident Command System provides the following five people as the staff for the incident commander. (See Figure 7.9.) Each of these individuals should have a maximum span of control of five.

1. Operations Officer
2. Planning Officer
3. Logistics Officer

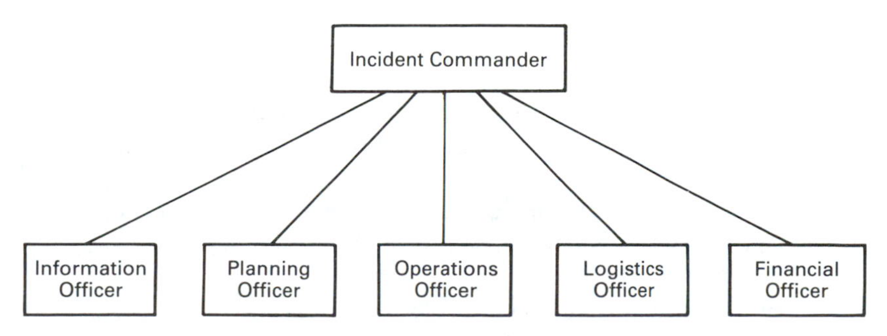

Figure 7.9 The incident commander's span of control

4. Information Officer
5. Finance Officer

The operations officer. The operations officer is responsible for all firefighting activities. His or her command is broken down into two or more divisions, depending on the magnitude of the problem. If air operations are included in the firefighting plans, an air operations officer would be appointed to coordinate these activities. This individual would also report to the operations officer.

The planning officer. The planning officer is responsible for assisting the incident commander in the preparation of the overall strategy for fire control. He or she keeps track of all the resources available and is responsible for keeping the incident commander up to date on the progress of operations. Any technical specialists assigned to the emergency would be under the planning officer's control.

The logistics officer. The logistics officer is responsible for coordinating support operations. He or she organizes the command post operations, establishes a staging area, and is responsible for feeding personnel and refueling apparatus, if required. He or she is also responsible for all emergency care rendered at the emergency and for maintaining the communication system. The logistics officer is assigned assistants in each of these areas, if required.

The information officer. The information officer acts as the link between the incident commander and the media. He or she is also responsible for maintaining liaison with other agencies such as the police department, Red Cross, and so on. His or her duties include monitoring and assessing hazardous or unsafe conditions and maintaining a high degree of safety for personnel. The information officer may also be assigned assistants if the magnitude of the situation warrants it.

The finance officer. The finance officer is responsible for all the financial aspects of the emergency. He or she is the liaison link between the incident commander and private contractors.

REVIEW QUESTIONS

1. Why is it important for a firefighter to have a basic understanding of the thought process the company commander goes through as he or she makes a decision as to the action to take at a fire?
2. What portion of the fires to which a company responds occur in target hazards?
3. What should a company commander do prior to leaving quarters after receiving an alarm of fire?
4. What are some of the thoughts that go through a company commander's mind as he or she responds to an alarm?

5. What are the three possibilities that the company commander will observe as the apparatus approaches the last available hydrant?
6. What action should the company commander take if nothing is showing?
7. What should the second company commander to arrive do when upon hearing the report to the dispatch center that nothing is showing?
8. What should the firefighter who goes into the building with the company commander to investigate take with him or her?
9. What action should the second company commander take if he or she sees the first company take in a line when the report was that nothing was showing?
10. What action should the first arriving company commander take if a small amount of smoke is showing on arrival?
11. What action should the first arriving company commander take if there are definite signs of a fairly good fire when he or she arrives at the last hydrant?
12. What action should the first arriving officer take when he or she arrives in regard to assuming command of the fire?
13. What are the three basic capabilities of an engine company in regard to handling lines?
14. How many additional engine companies should the first arriving officer request if he or she feels three more companies are needed to man 2½-inch lines for fire attack, two more for protecting exposures, and one more to man a portable deluge set?
15. What guidelines can be used for requesting additional truck companies?
16. What guidelines can be used for requesting salvage companies or companies to do salvage work?
17. When should emergency medical care capabilities be requested at a fire?
18. When is coordination required at a fire?
19. Is it fairly standard procedure that the first arriving officer actually takes command of the emergency?
20. What are the five things that an officer takes full responsibility for when he or she assumes command of the fire?
21. What is the next step that the officer in command should take after requesting additional assistance?
22. How should the orders directing the placement of companies be worded?
23. What things should be done at the change of commands?
24. For what purpose has the Incident Command System been designed?
25. What should be the maximum span of control for officers at a fire?
26. If a division chief takes command of a fire situation in which two district chiefs are assistants, where should the division chief normally place the district chiefs at a building fire?
27. Who are the five officers that normally report to the incident commander at a large scale emergency?
28. For what operations is the operations officer responsible?
29. For what operations is the planning officer responsible?
30. For what operations is the logistics officer responsible?
31. For what operations is the information officer responsible?
32. For what operations is the finance officer responsible?

CHAPTER 8

Fires in Buildings

OBJECTIVE

The objective of this chapter is to introduce the reader to firefighting tactics as they apply to building fires. Attic fires, basement fires, chimney fires, partition fires, dwelling fires, garage fires, mercantile fires, industrial fires, fires in places of public assembly, taxpayer fires, church fires, fires in buildings under construction, and high-rise fires are explored. Review questions are provided at the end of the chapter to assist the reader in determining his or her level of comprehension of the chapter contents.

As mentioned in previous chapters, no two fires are exactly alike; however, all fires are somewhat alike. Their similarity is the cornerstone for the establishment of the basic tactics of locating, confining, and extinguishing. Their differences are the foundation for the requirement of having to know more than the basics in order to become effective in fighting fires.

The location of the fire and the type of occupancy in which it is found are two of the many factors that contribute to the differences between fires. This chapter will attempt to highlight some of the variables found in various types of buildings and occupancies that influence the fire problem.

Prior to discussing fires in different locations and different types of occupancies, it is well to review several important points that apply to all types of fires in buildings. First, it is worth keeping in mind that fires in buildings cannot be put out from the sidewalk until the building has collapsed and has become one big bonfire. It is necessary to go inside and put water on the seat of the fire. The quicker this is done, the sooner the fire will go out. Of course, if the building is in danger of collapse, then safety of personnel requires that the fire be fought from the outside. There is no building worth the life of a firefighter.

If it is possible to go inside to work on the fire, it will require that personnel move into locations where it is hot and smoky, sometimes so smoky that the firefighters cannot see their hands before their face. These conditions make the use of breathing apparatus a must. Not only will breathing apparatus permit a more aggressive attack on the fire but they will also permit personnel to operate with greater safety.

A majority of all fires in buildings require only a small amount of water for extinguishment. Small lines are easy to maneuver and are normally effective for making a quick knockdown. It is important to remember, however, that a backup line should always be laid from the hydrant to ensure an ample amount of water whenever a small line is used for the initial attack in a building.

Another good practice at building fires is to take advantage of whatever built-in protection is provided. Interior standpipe lines can effectively be used to hold a fire in check while fire department hose lines are being laid. Lines laid into the exterior dry standpipe system will make water available at every floor and eliminate the need to lay lines up fire escapes or weave them up interior stairways. Lines laid into a sprinkler system will ensure an adequate supply of water on the fire or will hold the fire in check until department lines can be brought into play. Lines laid into the sprinkler system will be particularly effective at basement fires.

ATTIC FIRES

Attic fires should be attacked as quickly as possible with the use of 1½- or 1¾-inch lines equipped with fog or spray nozzles. Fog nozzles are preferred. The lines should be taken inside the building and advanced to the attic through the scuttle hole or attic stairway. (See Figure 8.1.) If neither of these openings is available or cannot be readily found, then the ceiling should be opened from below and a ladder extended through the opening provided.

As little water as possible should be used to control the fire. If possible, salvage operations should be started on the floor below the fire simultaneously with the line advancement into the attic. If water is used sparsely, there should be ample time available to spread covers before the water comes through the ceiling. If the attic is unfinished, sawdust placed between the joists may be effective in holding the water to the attic area.

It may not be necessary to open the roof for ventilation if the fire is small. The roof should normally be opened for larger fires to help confine the fire and enable the hosemen to gain access into the attic. Except in extreme cases, lines should not be directed through the hole in the roof. This will drive the fire back into the faces of the hosemen advancing the line. If a line is taken onto the roof it should only be used to protect the exposures or knock down any fire that spreads to the roof covering.

Gaining access to the attic from below is sometimes difficult in large finished attics with gable or arch roofs. Effective work can be done in these cases by opening the roof near the peak for ventilation and then making an opening near the eaves for the purpose of advancing a line.

Any loose shingles or roof covering should be removed during the overhaul process. A thorough check should be made for any downward extension of the fire. Particular care should be taken if hanging ceilings are involved. The hole in the roof

Figure 8.1 Attack attic fires through the scuttle hole

should be covered during the overhaul process if there is any chance that the weather may cause further damage.

BASEMENT FIRES

Probably more basement fires get out of control because of poor firefighting procedures than any other type of fire. This is primarily due to the complexity of the problem. All types of material can be found stored in basements. Fires starting in basements generally burn slowly and many times will smolder over a long period of time before being discovered. Because of the limited supply of oxygen in the basement area, they will give off large quantities of heat and smoke, which will not only contribute to the spread of the fire but will make access to the fire difficult. It is possible for the fire to spread very rapidly out of the basement through various vertical openings such as stairways, air conditioning systems, laundry chutes, and wall partitions. Sometimes these vertical openings, which are not commonly known to firefighters, are readily apparent on a visual inspection.

One of the first things that should be done upon arrival at a basement fire is to make a survey to see if heat or smoke is spreading to the upper floors or to the attic through any of the vertical openings. The building should always be evacuated as a precautionary measure against this possible spread.

The first lines should be brought into the ground floor and used to prevent any spread up the vertical openings. Spray nozzles should be directed into the vertical openings, and a check should be made to see if any fire is running the partitions. While operations are being conducted to control the upward spread of the fire, plans should be made as to the best method of attacking the fire.

The method of attack will depend on the number and types of access openings into the basement. The task is easiest when there are openings of sufficient size at opposite ends of the basement. Once lines are in position for the attack, the access on the leeward side of the building should be opened, followed by opening the access on the windward side. (See Figure 8.2.) Lines are advanced down into the basement from the windward access and the fire and heated gases pushed out the other end. Several lines should be worked abreast if the fire is of any size. It is important that lines be positioned on the leeward side to protect the exposures as the fire and heat is pushed out. In basements that have two openings as just described, never bring lines through both openings. This would result in each of the advancing crews fight-

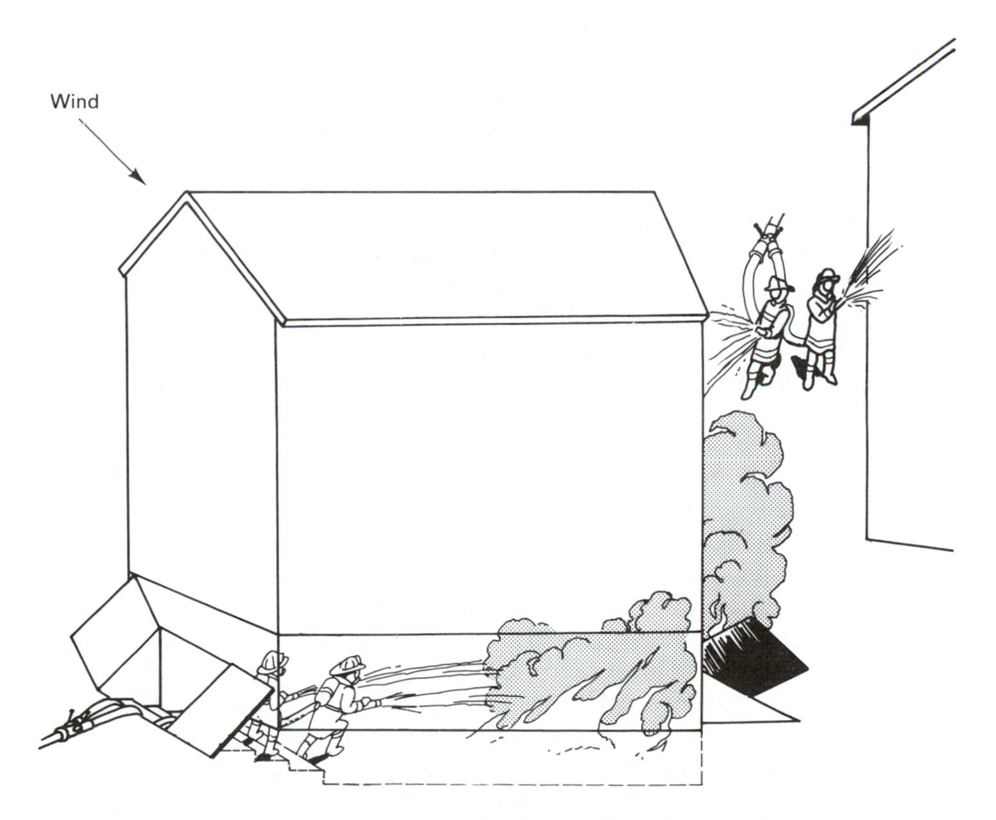

Figure 8.2 The proper method for attacking a basement fire

ing one another and would make the entire area untenable, as the heat and gases would have no place to escape.

The problem is much more complex when there is only one access to the basement area. Sometimes there are glass blocks in the sidewalk that can be broken to provide a second opening; if not, it may be necessary to make an opening in the floor above the basement. If this becomes necessary, select a location at the opposite end of the basement from the access. The opening should be near a window leading to the outside. The first floor should be cross ventilated and lines be placed into position at the location where the hole will be made. These lines will be used to push the heated gases and smoke out the window as the hole is cut. Once the hole is cut, lines can be advanced into the basement through the access area.

It may not be possible to advance lines into the area if the basement is well involved with fire. In this case it will be necessary to effect extinguishment by the use of cellar pipes and circulating nozzles. Several holes will have to be cut in the floor above the fire. Where possible, holes should be cut over the hottest part of the fire. Many times these locations can be determined by feeling the floor with the hand. Care should be taken to have protective lines in place as the holes are cut, as heat and gases can be expected to be discharged from the cut holes. Every effort should be made to keep the heat and gases in the basement area.

If the basement is well involved with fire there is a good possibility that the floor above the basement can become so weakened by the fire that collapse is possible. Therefore, it is good practice to remove all personnel from the building once cellar devices are discharging water into the fire area. Of course, it goes without saying that all personnel should be removed from the building any time the floor gives any indication that it is weakening to the point that it might collapse, regardless of whether or not water is being directed onto the fire. In these cases it will be necessary to do whatever is possible to stop the fire from outside of the building.

When cellar appliances are used to extinguish the fire, their use should be discontinued once access to the basement can be made with handlines.

Sometimes fires occur in basements that are protected by automatic sprinkler systems. In such cases the fire usually does not make much headway unless it is traveling in concealed spaces that are inaccessible to the flow of water from the sprinkler system. When fires occur in sprinklered buildings, the sprinkler system should be supplemented by department pumpers. The system should be kept in operation until lines are in place and there is assurance that the fire will not spread once the sprinkler system is shut down.

Although all departments do not have the capability of providing it, high expansion foam has proven to be beneficial in the control of basement fires. It expands at a rate of 1,000 (or more) to 1 and will completely fill a basement in a short period of time. It has the capacity to smother the fire, is nontoxic, and leaves no residue. It is good practice to use it if it is available, particularly in those cases where access to the basement area is limited. It is particularly effective in small basements.

As a last resort the basement can be completely flooded. At some basement fires it may be the only method of gaining complete control.

A couple of last thoughts on basement fires are noteworthy. Basement fires are death traps. If a fire has been smoldering any length of time in a basement, conditions are almost certain to be ripe for a backdraft. Firefighters have been cremated in basements when the area suddenly got an unexpected supply of oxygen. The backdraft potential should always be eliminated before personnel are sent into the area. Another hazard to personnel at basement fires is the utilities. It is not unusual to find high-voltage electrical equipment or exposed gas lines in the basement. These may become extreme hazards if exposed to water or fire. It is therefore imperative to shut off all gas and electrical lines to the building prior to attacking fires of any size in the basement area.

CHIMNEY FIRES

Fires in chimneys and flues usually do not present too much of a problem, but they should not be taken too lightly. The problem usually is not in the extinguishing of the fire but in making sure that the fire has not extended into hidden portions of the building.

If the fire is contained within the chimney it is best to allow it to burn itself out. If it is necessary to use water, it should be used very sparsely so as not to crack the masonry. A small spray from a garden hose nozzle is usually sufficient. If the chimney is stopped up, it will be necessary to use a ball and chain or some other weighted object to create a clearance. The ball or weighted object should be wrapped in burlap, tied to a rope and then worked up and down in the chimney until the obstruction has been cleared. A salvage cover or something similar should be used to cover the fireplace opening prior to commencing this operation. This will prevent soot or pieces of the obstructing material from getting into the room.

A check should always be made to ensure that the fire has not extended into other parts of the building. It will be necessary to examine all intervening floors if the chimney extends from the first floor to the top of the building. The attic should also be examined to ensure that the fire has not extended into this area by conduction or through cracks in the bricks. Smoke in the attic or in other portions of the building from a chimney fire is a pretty good indication that there are cracks in the chimney lining.

PARTITION FIRES

Partition fires are often overlooked and therefore are a primary cause of rekindles in building fires. They are dangerous from the standpoint that they normally do not display themselves until considerable damage has been done. The possibility of partition fires is more prevalent in older buildings where wood lath has been used in the wall construction and there is an absence of fire stopping. The lath becomes dry and the roughness can hold small sparks of fire over a long period of time.

One of the best visual indications of a partition fire is discoloration of the wall paper or paint, but this sign is seldom evident. The best way to detect these hidden fires is to feel the walls carefully with the hand. If the wall is too hot to touch there is probably a fire inside. The walls on the fire floor, the floor below the fire, and the floor above the fire should be checked for partition fires. It is particularly important to check the floor above if the fire originated in the basement.

It is necessary to open up the wall whenever indications of a possible partition fire have been found. The wall should first be opened between the studs near the baseboard. A small spray of water should then be directed into the opening if any signs of fire are found. The stream should be directed upward to stop any extension of the fire and the spray allowed to trickle down the insides of the wall to extinguish any fire found there. The wall should then be further opened between the studs until all the burned area has been exposed.

Some departments are reluctant to open up walls to check for partition fires. It is true that there are certain types of plaster walls that hold the heat for a considerable period of time and can give indications of a possible partition fire. There is no doubt that many walls have been opened up and no evidence of fire found inside. However, the expense of the repair is minor compared with what could happen if a fire in a partition is allowed to go undetected.

DWELLING FIRES

Dwelling occupancies vary from small, one-story, single-family residences to multistory tenements, apartment houses, and hotels. Construction may be wood, stucco, brick, or various types of masonry. Some are considered as fast burners, others are classified as fire resistive. Although older structures are generally considered more of a fire problem, some of the newer, larger dwellings with their high ceilings and

Photo 8.1 Well-involved dwellings are not uncommon. (Courtesy of Chris Mickal, New Orleans Fire Department Photo Unit)

open spaces provide the setting for a rapid spread of the fire and a greater potential for a larger loss. For example, the estimated fire loss for the building and contents of a single-family dwelling which occurred in Malibu, California, in 1989 was $7.5 million. This is nearly six times more than the entire fire loss for the conflagration that swept through Chelsea, Massachusetts, in 1973 in which 300 buildings were involved. The trend toward larger and more expensive homes, primarily those that are two stories or more in height, has not lessened the fire problem in dwellings.

Dwelling fires require the employment of almost all the basic firefighting tactics that were discussed in previous chapters. Of course, the primary problem is the life hazard. This is apparent when one considers the fact that a large majority of all those who die in fires die where they live. Of these, most die during the sleeping hours. This does not mean, however, that the hazard does not exist during daylight hours. The life hazard is always present and should be given first priority at every fire. An immediate search should be made of all smoke-filled areas, with the search beginning as close to the fire area as possible. Ladders will be required for evacuation of upper floors of larger occupancies, and it will be necessary to search every room. Although the life hazard takes first priority, it should be kept in mind that in a majority of cases the most effective means of saving life is an aggressive attack on the fire.

Single-Story Dwelling Fires

Fires in one-story residences vary from those that can be put out with a fire extinguisher to those where the entire building is well involved with fire. Single-room

Photo 8.2 The primary objective in dwelling fires of this magnitude is to protect the exposures. (Courtesy of Chris Mickal, New Orleans Fire Department Photo Unit)

fires, however, are much more common than well involved dwellings. Single-room fires can generally be extinguished with an 1½- or 1¾-inch line, backed up by a larger line. The line is brought in from the uninvolved portion of the dwelling with the objective of confining the fire to the single room.

Well-involved, one-story dwellings of average size can generally be attacked with the use of 1½- or 1¾-inch lines. The primary objective with this type of fire is to keep the exposures from becoming involved. In some parts of the country dwellings of this size are built fairly close together. If the siding is wood, the exposure problem can be severe. The normal procedure for this type of fire is for the first arriving engine company to make an attack with two 1½- or 1¾-inch lines. The 1½- or 1¾-inch lines are used to protect the exposures on either side of the fire, with the first line going to the leeward side. The actual attack on the fire is made by later arriving companies.

The heat given off by larger one-story dwellings can be severe due to the tremendous amount of combustibles involved. Fortunately, there is generally a greater distance between dwellings of this size, which reduces the exposure problem. However, if the fire building is equipped with wood shingle roofs, consideration should always be given to the possibility of flying brands starting roof fires downwind. In this case it is advisable to have an engine company patrol downwind as a safety factor against this threat.

It is good practice to commence the attack using 2- or 2½-inch lines equipped with spray nozzles. Few of these larger fires can be knocked down by less than two of these larger lines. The 2- or 2½-inch lines should be reduced to 1½- or 1¾-inch lines once the fire has been knocked down to manageable size. The 1½- and 1¾-inch lines are more maneuverable and their employment releases personnel that can be used elsewhere on the fire.

Fires in attics or basements of dwellings should be handled as discussed under these separate headings. Attic fires require a coordinated salvage effort as previously discussed in Chapter 5.

Multiple-Story Dwelling Fires

Firefighters most fear fires of any size in multiple-story dwellings. This is particularly true when the fires occur at night, primarily during sleeping hours. There is always the life hazard and the constant threat to firefighters in their attempt to make rescues and carry out their assigned functions.

Fires may vary from those in two-story, single-family dwellings to larger hotels housing hundreds of transients. The most threatening fires are probably found in the older, multistory buildings with their wooden construction, open stairwells, and other vertical openings, with the resultant constant threat of rapid fire spread. Buildings of complete wooden construction are generally limited to four stories in height, but so-called tenement buildings having wooden doors, floors, beams, and stairways with brick walls extend upward to seven stories.

Rescue, salvage, and ventilation should be carried out at these fires in accordance with the procedures earlier established. These functions should be coordinated with a rapid and aggressive attack on the fire.

Photo 8.3 Fires in multiple-story buildings may require both outside heavy streams and inside hose lines. (Courtesy of Chicago Fire Department)

The first lines should simultaneously be advanced to the fire floor (or floors) and the floor above the fire. Lines can be taken to the floor above the fire by way of the fire escape or ladders. The inside stairway can also be used if the fire is not blocking this pathway. Extreme care must be taken to ensure that an escape pathway is maintained for the line taken above the fire. This line should be used to gain control of the vertical openings and hold the fire to the fire floor. Lines can be advanced to the fire floor by the same pathways as used for taking the line above the fire floor. Where possible, this line should be advanced onto the fire from the noninvolved portion of the building to keep the fire from spreading horizontally. The top floor should be checked as soon as possible to ensure that the fire has not spread to this area or the cockloft before lines were brought into position to stop the vertical spread.

If large volumes of fire are visible upon arrival, it might be wise to use heavy streams to knock down as much fire as possible while handlines are advanced into position. These lines should be shut down, however, once handlines have been taken into the building.

GARAGES

Extinguishing a fire in a garage appears to be a simple process—hardly worth discussing in a training manual on firefighting. There are a couple of problems that are worthy of consideration, however.

Garages in single-family dwellings are increasing in size. One-car garages were once considered as standard, but today three-car garages are no longer considered an exception. Their locations have also changed. Attached garages have almost become a standard, whereas once they were quite rare. The increase in the size of the garage and the change in the location has drastically increased both the attack and the exposure problem.

The exposure problem from a well-involved detached garage fire is generally solved by advancing an 1½- or 1¾-inch line between the garage and the house and keeping the house wetted down. The exposure problem from an attached garage might involve taking a line both inside the house and working another outside the house.

The attack on the fire itself is generally made by advancing an 1½- or 1¾-inch line to the front or side door of the garage. Some departments make it standard practice to force up the overhead garage door, and make a direct attack on the fire. The door is normally held open by springs; however, these springs should not be depended on during firefighting operations. If the springs should break, the door will come down in a hurry, possibly injuring firefighters or trapping them inside the garage. Consequently, standard operations should require that the door be propped open with the use of a pike pole or similar object. The line that will be used for the attack should be used to protect the firefighters opening the door until it is secured in place.

Attached garages may be found next to or under both one-story and two-story residences. Those next to a one-story dwelling will require that a line be taken inside the house to prevent extension of the fire into this area. A 1½- or 1¾-inch line will generally be sufficient for this purpose. A two-story residence with the garage attached at the first floor and under the second floor will require that two lines be taken inside. One-inch lines will generally suffice; however, if a stairway leads from the garage to the second floor, a 1½- or 1¾-inch line should be used, as the chance of the fire extending into the house at this location is greatly increased. A 1½- or 1¾-inch line should also be used at the top of the stairway in those dwellings where the garage is under the house with a stairway leading up to the house. (See Figure 8.3.)

Extreme care should be used when making the attack on the main body of fire. Breathing apparatus must always be worn and a backup line should always be provided for the main attack line. This is necessary because it is impossible to know in advance what type and amounts of material will be involved in the fire. In addition to ordinary combustible materials, garages can be expected to act as a storage location for gasoline, kerosene, pesticides, fungicides, propane tanks, spray cans, and other materials that could become an extreme life and health hazard to firefighters. Every effort should be made to keep any containers identified as containing gasoline, kerosene, propane, or similar material cooled down during the entire firefighting process. If a hazardous material team is available, it is a good idea to have its members evaluate the seriousness of the material encountered.

It can also be expected that magnesium may become involved, for it is not unusual to find wheels on lawn mowers and motorcycles made of this material. Its involvement would be indicated by a brilliant white light in the area of its burning.

Figure 8.3 Prevent the fire from spreading into the dwelling

Care should be taken when directing water into this area, as small explosions can be expected when the water comes into contact with the burning magnesium. A good strategy to adopt when fighting fires in garages is always to expect the unexpected.

MERCANTILE FIRES

Mercantile establishments vary from those that are relatively small to some that are extremely large. The typical store has a large undivided sales area in the front and a small service area in the rear. Most fires start in the service area. If the fire has not progressed into the sales area by the time the department arrives, it can usually be confined to the rear of the building by bringing lines in through the front and making an aggressive attack on the fire.

Life safety can be a problem if the fire occurs during the hours the business is open. Most shoppers will attempt to leave the store through the entrance they

Photo 8.4 Fires in mercantile occupancies are common in many cities. (Courtesy of Wichita, Kansas, Fire Department)

came in. This is usually in the front of the building—where the department will be bringing in lines. Large show windows are usually found in the front of the occupancy. If necessary, all of these can be removed for access and ventilation.

The potential for a fast-moving fire generally exists in mercantile establishments due to the large quantities of burnable material found in the sales area. There is also a potential for a large loss due to the value of the merchandise on display. Every effort should be made to commence salvage operations immediately. Care should be taken during the overhaul process to collect burned stock in such a manner that as accurate an inventory as possible can be made of the damaged material.

Smoke can cause a considerable loss to clothes and similar items. Ventilation should be started early and made in such a manner as to keep smoke out of the sales area, if possible. Roof ventilation is usually preferred. As many fans as possible should be set up to create a positive pressure atmosphere.

INDUSTRIAL FIRES

Serious fires in industrial plants are less frequent than in mercantile establishments due to a fire-conscious management and better built-in protection. The potential loss, however, is severe due to the high values involved. Industrial plants generally contain large open areas, which allow rapid extension of the fire once it gains any headway whatsoever. Stopping the spread and extinguishing the fire presents some serious problems. Many times it is necessary to make the initial attack using heavy stream appliances until handlines can be brought into position. If the fire has gained

Photo 8.5 Warehouse fires can present some serious problems. (Courtesy of Chris Mickal, New Orleans Fire Department Photo Unit)

considerable headway it may be necessary to continue the use of heavy streams due to the limited reach of handheld streams.

Firefighting tactics in large industrial fires consist of protecting the exposures and extinguishing the fire using as many lines and firefighters as necessary. There are, however, some problems peculiar to industrial plants that should be considered when smaller fires are encountered, or in the early stages of larger fires.

Large quantities of flammable liquids in containers may be found in some occupancies. Every effort should be made to keep the containers cool to prevent their rupture or explosion and the subsequent rapid spread of the fire. Spray streams are particularly useful for this purpose. (See Figure 8.4.)

Every effort should be made to keep acids or other corrosive materials from coming into contact with combustible material. If acids and cyanides are stored or used on the premises, they must be kept apart at all costs. (See Figure 8.5.) Mingling of the two will result in the formation of hydrocyanic acid gas. This gas is so powerful that one breath can result in death. Breathing apparatus do not provide protection against it, as the gas is absorbed through the skin. These materials may be found in metalworking establishments. A hazardous material team should always be requested, if available, whenever these materials are encountered.

Fires in areas where molten metals are found should be extinguished using high-pressure fog streams. Solid streams played on molten metals could result in violent explosions and splattering of the metal.

Fires in large sawdust or dust-collecting bins should be initially knocked down

Figure 8.4 Protect exposed flammable liquid containers

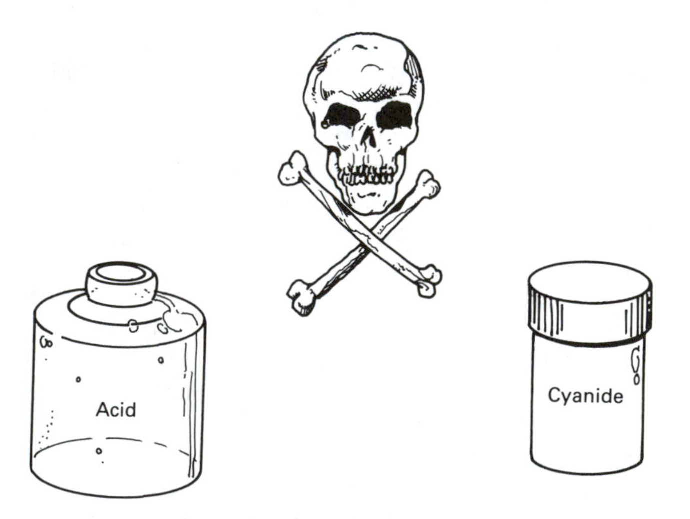

Figure 8.5 This combination is a *killer*

using spray streams. After the fire appears to be extinguished, the contents of the bin should be dumped through spray streams to prevent a possible dust explosion. (See Figure 8.6.)

Vats of hot salt baths present a similar problem to that of molten metals. If possible, the vats should be covered prior to extinguishing any fire around them. If the vats cannot be covered, the fire should be knocked down by using high-pressure fog nozzles.

Fires in dip tanks or large vats of combustible paints can be extinguished by the use of water, but care must be taken to not overflow the container. This can generally be prevented by using high-pressure fog nozzles for extinguishment.

Fires in lumberyards or in the lumber storage area of industrial plants frequently require heavy streams for extinguishment. This is due not only to the size of the fire but the tremendous heat that is being given off. The heat may be of sufficient intensity that handlines cannot be advanced close enough to spray water on the fire. There is almost always the hazard of flying brands from this type of fire, which requires that the area downwind from the fire be patrolled.

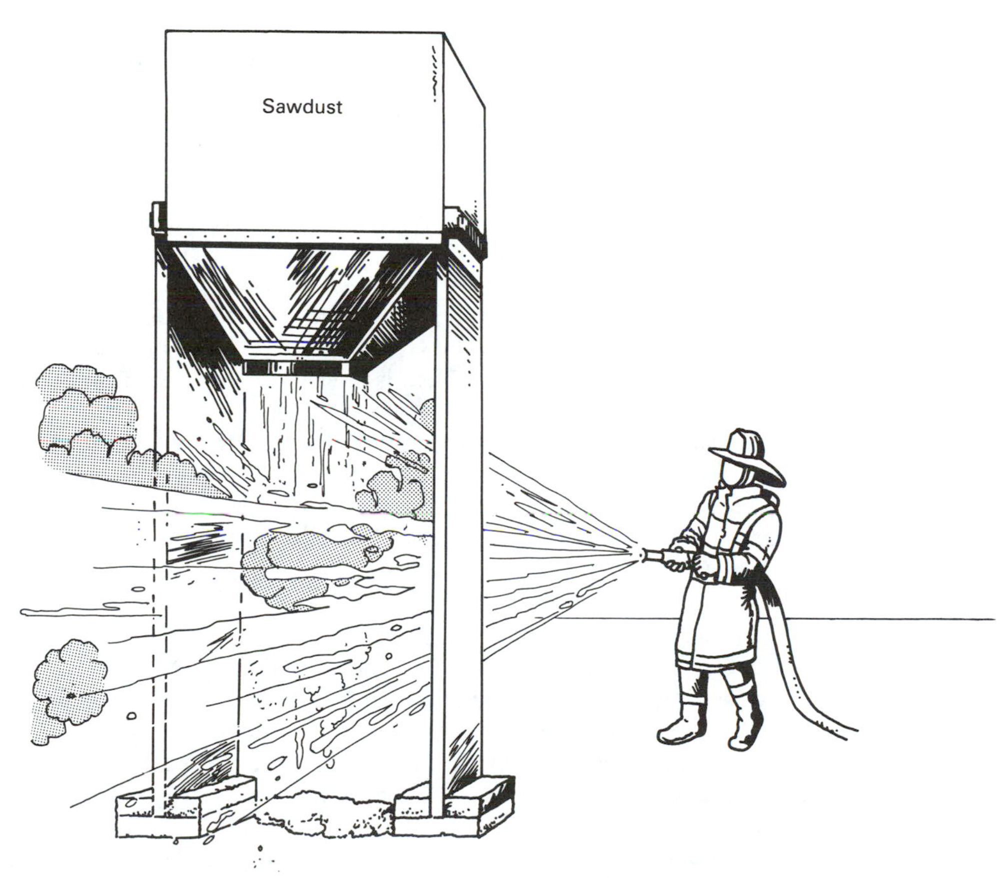

Figure 8.6 Dump the sawdust through a fog or spray stream

PLACES OF PUBLIC ASSEMBLY

Some of the most disastrous fires from the standpoint of loss of life have occurred in places of public assembly. Three of the most notable were the Iroquois Theater fire in Chicago in 1903 (602 dead), the Cocoanut Grove fire in Boston in 1942 (492 dead), and the Beverly Hills Supper Club fire in Southgate, Kentucky, in 1977 (165 dead). Panic was a big contributor to the loss of life in each of these fires. Thus, it therefore follows that one of the primary objectives of the fire department at public assembly fires is to prevent or eliminate panic.

Panic is the result of fear. A single shout of ''Fire!'' can convert a calm group of people into a raging mob. The sound of a siren may possibly accomplish the same result. Apparatus operators responding to a reported fire in a place of public assembly should use due caution with the use of the siren. It is wise to cut off the siren if nothing is showing as the reported location is approached. Of course if there are signs of an active fire, operations should proceed as needed, with the expectation that panic has already become a part of the exit process.

One of the first actions that should take place upon arrival is to see that all exits are thrown open immediately. Lines should be advanced to protect the exits and ladders raised to every fire escape and every window where people are showing, as an assist to emptying the building as quickly as possible. Ventilation should be started to pull the smoke and heat from the exit pathways and to help relieve the panic-contributing atmosphere that exists inside. Ambulances should be requested, as they will no doubt be needed.

The first lines should be used to protect the exits until everyone is safely out of the building. It may be that the best method of protecting the exits is by a rapid attack on the fire. This would certainly be the case if the fire was threatening to move in a direction that would cut off exit pathways.

A thorough search of the building should be started as soon as possible after arrival to ensure that no one has been trapped or overcome by smoke. It should never be assumed that the rescue problem is over when people quit moving out of the exits.

Thought should be given early in the emergency to the possibility that the fire will involve the electrical equipment and throw the building into darkness. This would not only increase the problem of rescue but would incite the crowd into a panic situation if such condition did not already exist. Consequently, portable lighting equipment should be taken into the building and plans made for its use, even if it appears that lighting will not be a problem.

Theaters are better prepared to cope with a fire situation than other places of public assembly such as nightclubs, dance halls, and hotel ballrooms. The most likely place for a fire to start in these occupancies is back stage. The stage area is designed with a vent that should open automatically in case of fire. A fire of any size should also cause the fusible link on the fire curtain to part, which will cause the curtain to drop. The fire curtain is designed to prevent fire, heat and smoke from entering the auditorium area.

When responding to a back-stage fire, the officer in charge should ensure that the curtain has dropped and the vent is opened. Lines should be brought into the auditorium to prevent extension of the fire into this area. Additional lines should be brought into the stage area from both sides if possible. Although theaters are generally made of fire-resistive material, there is normally sufficient combustible material back stage to generate a good-sized fire.

TAXPAYER FIRES

The term *taxpayer* came about in the early 1900s to identify a building that was constructed for the purpose of saving money. An investor would buy a piece of vacant land that he or she anticipated would increase sharply in value within a few years. The individual would then construct a building as cheaply as possible to provide sufficient income to pay the taxes during the interim required for the land to increase in value. The term has been expanded to incorporate a type of building with common problems. The building is normally one or two stories in height and contains a number of small businesses under a common roof. The primary feature is an undivided attic or an occasional dividing wall of weak construction. If the building has a basement, it will also normally be undivided, with wire or flimsy walls separating the storage area of one business from another.

The type of construction and the contents of the various businesses provide the potential for a rapid spread of fire; however, the primary problem is the undivided attic and possibly undivided basement. The key to fire control is to get ahead of the fire and stay ahead of it. Of course this is a basic principle at all fires, but with this type of fire failure to attack rapidly and aggressively will probably result in the loss of the building.

Upon arrival it is sometimes difficult to determine exactly which store is involved. Smoke will fill the attic space and will be pushing out from all parts of the building. Visible flames help in locating the fire. There are several actions that should be taken once the location of the fire has been identified. A large hole should be cut in the roof directly over the fire. It is good practice to make this hole at least 10′ by 10′ in size. Simultaneously, a hole should be cut in the ceiling of the second store on either side of the involved area. Ladders should be placed into the openings and lines advanced into the attic to stop the spread. Lines of 1½ or 1¾ inches will normally be sufficient for this purpose. (See Figure 8.7.)

The problem will be increased if the building is two stories in height. The same general attack procedure will prevail if the fire is in a store on the second floor. However, the salvage problem has been intensified. Merchandise in the store below the involved area should be covered first, with salvage operations continuing in the adjacent stores as soon as possible.

If the fire is of any size in the basement area it can be anticipated that the entire basement area will probably be lost. Standard procedures for extinguishing fires in basements should be employed, with the hopes that the fire can be confined to the basement area.

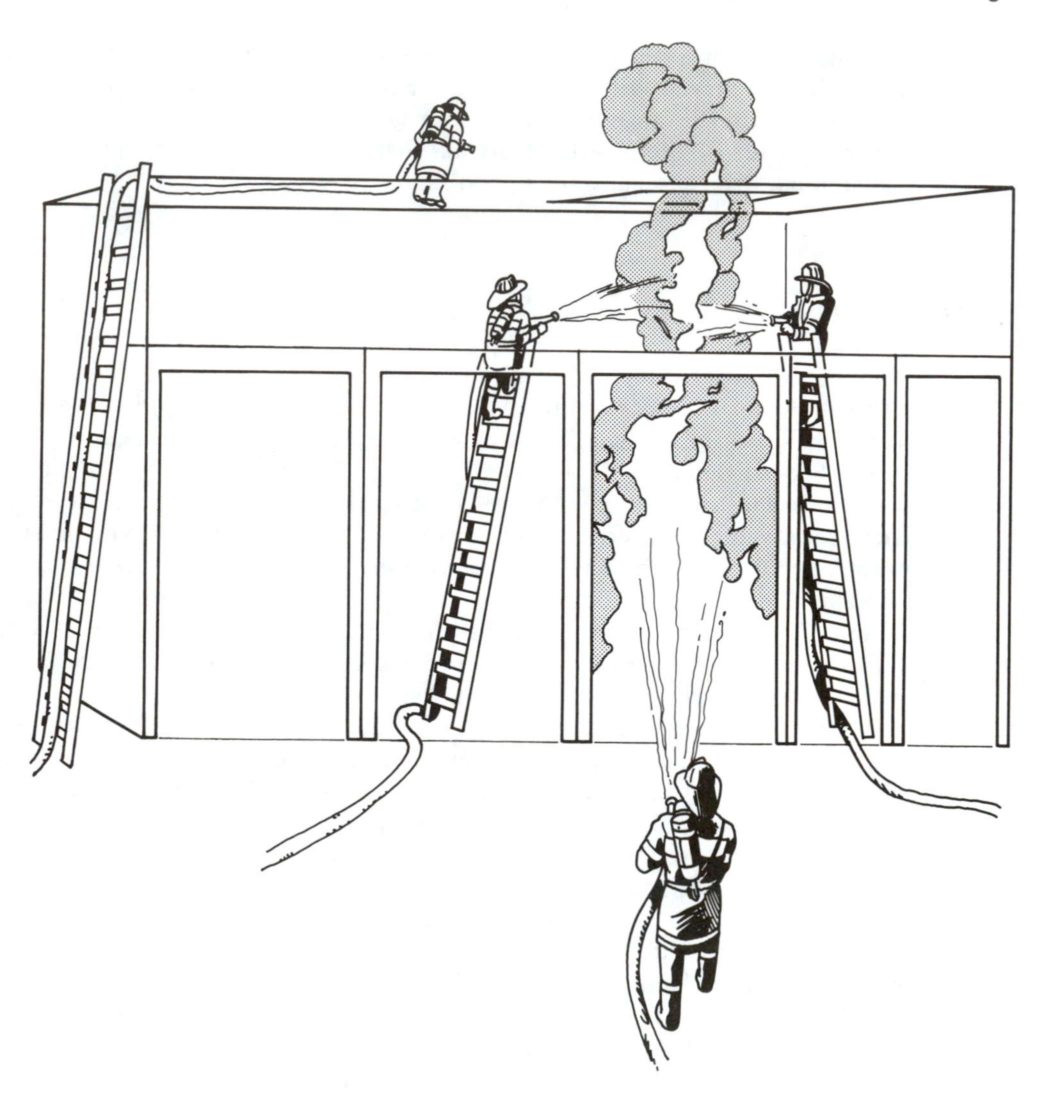

Figure 8.7 Stop the spread and extinguish the fire

CHURCH FIRES

Church fires are different from almost any other type of fire to which the fire department may respond—different from the standpoint that it can almost always be anticipated that if the fire is of any size at all, the entire building will probably be lost. The primary reason for this is the construction of the church itself. The interiors of most churches, and in particular the older ones, are made of wood, with wood also used in the pews and benches. Wooden choirlofts and wooden balconies are not uncommon. There is usually a high-pitched roof with a large-vaulted ceiling and a steeple. The ceiling is normally too high to be opened if fire invades that area.

Photo 8.6 Church fires present unique problems to firefighters. (Courtesy of Chicago Fire Department)

The wide open spaces within the church create drafts that have the potential of whipping fire in all directions.

Concealed spaces run rampant throughout a church. In addition to those in the vaulted ceiling and organ loft, many will be found in the walls. Additional construction is added to hide heating pipes, which run through the main auditorium as well as through rooms to the rear of the church and in the organ gallery area. Walls, recesses, and pipe channels are many times furred out to give a smooth appearance, which results in providing concealed spaces that run a foot or more in depth. Unfortunately, most of the furred areas terminate in the vaulted ceiling or the attic.

Most churches have basements. Statistics indicate that a large number of the fires in churches originate in this area. This might be contributed to the fact that the heating unit is normally located in the basement. A fire originating in this location has a good chance of spreading rapidly throughout the entire church. However, if the fire department receives an early call, it may be possible to hold the fire to the basement by a rapid attack on the fire. Concealed spaces in walls should be opened and lines used to prevent the upward spread. If these concealed spaces are not immediately controlled, it is likely that the fire will pass the auditorium area and extend directly into the huge hanging ceiling area or the attic. Once this happens, the building is lost.

Fortunately, most fires in churches start when the church is unoccupied. This reduces the life hazard but does not eliminate it entirely. Many of the churches have an unprotected passageway that extends from the church to a rectory, where it is common to find people at all hours. This passageway will permit fire to travel in either direction. Lines should be advanced into this passageway from the uninvolved area if a fire of any size is apparent in either the church or the rectory when the department arrives on the scene.

Fires that have gained any headway in a church should be attacked with the use of heavy lines. If the fire has entered the hanging ceiling or enclosed attic area, there is little chance of extinguishing it. The ceiling is too high to open and it is too hazardous to place firefighters on the steep roof to ventilate. Elevated platforms or ladder pipes should be set up to attack the fire once it burns through the roof; however, care should be taken as to where these apparatus are placed. If the fire has or is likely to extend into the steeple, it is almost a certainty that this portion of the church will eventually collapse. The steeple acts like a chimney flue and will quickly pull fire through its entire interior. Plans should be made for this collapse with thought given to the hazard created by the heavy bells that are located in the steeple area.

It is normally dangerous to operate inside the church if the fire is of any size whatsoever. The possibility of heavy lighting fixtures falling and the potential failure of balconies and organ loft structures create too high a risk for firefighters.

FIRES IN BUILDINGS UNDER CONSTRUCTION

A fire in any frame building under construction will present unusual problems to firefighters; however, the most troublesome appear to be those large complexes of two-story apartments, condominiums, or townhouses. Some of these developments extend over an area of several blocks. It is not unusual to find these complexes being built in the midst of a well-developed area of dwellings with wood shingle roofs. They are particularly vulnerable to rapid fire spread when they are in the framing stage prior to drywall or other covers being installed. A fire starting at this stage of construction can quickly become a two-story bonfire, with all the potential of developing into a major conflagration. This is particularly true during periods of low humidity and high winds.

Not only do fires in these complexes present a constant threat of rapid fire spread but firefighters are normally hampered by an inadequate water supply and an extremely difficult access problem. It is not uncommon to find these complexes in the framing stage prior to water mains being installed. Additionally, deep trenches, piles of dirt, and stacks of construction material may block apparatus access and hamper the advancement of hose lines. The problem can become acute right after a rain, which fills trenches with water and makes the entire area one big puddle of mud.

This description paints a pretty dismal picture but one that is factual in all respects. It illustrates the problems involved and offers no solution for effective firefighting operations. In reality, the only effective method of fighting a fire in these types of complexes is to start before the first nail is driven. This means that a city must pass adequate ordinances and ensure proper enforcement. No construction should be started until the water mains have been installed, the system tested, and the streets properly paved. On-site hydrants should be provided if the distance from street hydrants is excessive. Inspections should be made during the entire construction process to ensure that access pathways are kept clear and that unusual problems do not develop that would hamper firefighting operations.

This is the ideal; unfortunately, it is seldom achieved. Consequently, fire officers have to attack the fire using the resources they have and under the conditions they encounter. Additional help beyond the first alarm assignment will undoubtedly be needed if the fire has gained any headway whatsoever. The first arriving officer should not hesitate requesting assistance. In fact, no harm can be done if a second alarm is requested if any size of working fire is observed in the vicinity of the complex while apparatus are responding.

The primary problem will be the rapid fire spread, with the resultant threat to exposures. If the complex is isolated, the threat may be confined to that of flying brands. However, in most cases the threat will include that of flying brands along with the problem of protecting the exposures from the tremendous radiant heat that will develop. The exposure problem from radiant heat may be confined to other units in the complex or may include the threat to other dwellings surrounding the complex. The primary problems will exist downwind.

Exposures can be protected by using basic firefighting tactics. This type of fire will call for large lines and the employment of heavy streams. A sufficient number of companies should be requested to provide the fire streams required to protect the exposures and make an aggressive attack on the fire. The first streams should be directed on the exposures and should be heavy streams. Several companies should be patrolling downwind to protect exposures from flying brands. It may be necessary that water be relayed from some distance if an adequate supply is not available in or close to the complex. Consideration should be given for the additional companies that may be needed for relay purposes.

HIGH-RISE FIRES

High-rise is a term coined by the media to refer to those buildings that are reaching toward the heavens. A number of definitions have appeared since the initial use of the term. Some fire departments have defined it to mean any building over 75 feet in height, primarily for the purpose of controlling by ordinance the built-in fire protection during the construction of such buildings. Others have defined it to mean any building beyond the reach of the fire department's aerial equipment, or to put it another way—those buildings too high to rescue all occupants by the use of the department's aerial equipment. This definition included buildings of less height than the aerial equipment if the building was set back a considerable distance from the street, which restricted the use of the aerial equipment to its maximum height. Regardless of the number of definitions, it appears to be safe to say that a high-rise building is any multiple-story building that requires the use of high-rise firefighting tactics for effective extinguishment. Present high-rise tactics have been designed primarily to cope with fires in the newer, higher buildings. These tactics merge the Incident Command System, the experience gained over the years in fighting fires in multiple-story buildings, and the bitter lessons learned from the many disastrous high-rise fires that have occurred around the world. Much of the information contained in this section on high-rise fires is based on the system developed by the Los Angeles Fire Department.

Construction Features

According to almost any of the accepted definitions, high-rise buildings include many older multiple-story buildings having unprotected vertical openings in addition to the newer structures that are reaching 50 stories and more. Although the newer, taller buildings present an excessive number of problems, they do have the advantage of fire-resistive construction and many built-in fire protection aids for both the occupants and the fire department. For example, the high-rise ordinances of some cities require that the building be equipped with the following:

1. A sprinkler system throughout the building.

2. An on-site water supply system with pumps capable of delivering water to the top-most floor through a combination standpipe system. The outlets on the system are 2½ inches, which provides the capability for the fire department to take lines aloft and work directly off the system. The system includes fire department inlets, which make it possible for the fire department to help boost the pressure if necessary.

3. A local fire warning system that is actuated either manually, by smoke detectors, or by water-flow switches. Some ordinances require that the building be thrown into an alarm mode when the system is actuated. The alarm mode will sound a localized alarm, then shut down the heating, ventilation, and air conditioning systems on the fire floor and in some cases on adjacent floors. All stairwell doors will be automatically unlocked and the stairwells pressurized. If the smoke detector in the elevator vestibule is actuated, the elevators on the affected bank will be recalled to the ground floor.

4. A one-way communication system that permits fire department officials to communicate with occupants in all or any single public area of the building.

5. A building control station specifically designed for fire operations. It houses a fire alarm annunciator panel, the building communication system controls, an elevator status panel, and the smoke-handling controls.

6. At least two means of egress from each floor other than the elevators.

7. A fire department lockbox that contains keys to operate the automatic elevators and the fire alarm system, and to gain access to floors from the various stairwells.

8. An emergency power system.

9. An emergency helicopter landing pad on the roof.

Because of the type of construction found in high-rise buildings, there is no fire on record involving the newer high-rise buildings that resulted in any type of building collapse. There is an advantage in firefighting to know that the ceiling is not going to fall in or the floor collapse. It should be noted, however, that the superior qualities of fire-resistant construction found in high-rise buildings also cause a severe problem to firefighters. When an entire floor is involved, the contained heat creates an oven-like atmosphere. This generally subjects firefighters to severe punishment when making an attack.

There are two basic types of high-rise buildings that have been designed for human occupancy: residential and commercial. Hotels, apartment houses, condominiums, hospitals, and physical care facilities are included in the residential type. The commercial type is occupied by offices and many types of businesses.

Most residential high-rise buildings are characterized by center corridors and a large number of interior compartments; however, some are designed around the center-core concept. These buildings are normally occupied on a 24-hour basis.

Commercial high-rise buildings are characterized by a center-core construction with circuit corridors around the core. Many contain a large number of open spaces.

The center-core design concept features office areas or living areas surrounding a "core" containing stairwells, elevators, and utilities. This design presents an unusual problem to firefighters. Firefighters advancing a line onto a fire can suddenly find themselves trapped by the fire behind them which has pushed its way around the core. Protection against such a happening can be provided by advancing a line in the opposite direction of the original. This line should not be used, however, to attack the fire, as it would result in the two streams fighting one another. (See Figure 8.8.)

The population density is usually much greater in commercial buildings than in residential buildings, with the greatest concentration occurring during business hours. Some buildings contain 5,000 or more people during this period. It takes a considerable period of time to remove 5,000 people from a building via the stairways, even under ideal conditions.

One construction feature that can hamper firefighting operations is the manner in which the ceilings are designed. The ceiling serves as the return air plenum for the heating, ventilation, and air conditioning systems. It provides a hidden space where the fire can spread rapidly and go undetected. This area should be constantly checked when fighting a fire on any floor. (See Figure 8.9.) A pike pole or similar tool can be used to remove one of the panels. If such a tool is not available, a quick burst of a hose stream will normally do the job.

Firefighting Problems

Fires in high-rise buildings present some unique problems to the fire service. One of the primary problems is getting to the fire. If the fire is high in the building it may take a considerable time to reach it. This gives the fire additional time to grow in intensity and spread if avenues are available. It also means that planning ahead is a must. An example is a fire that occurred on the twenty-first floor of a building in one of the larger cities. Although the fire was well fought, 36 minutes elapsed

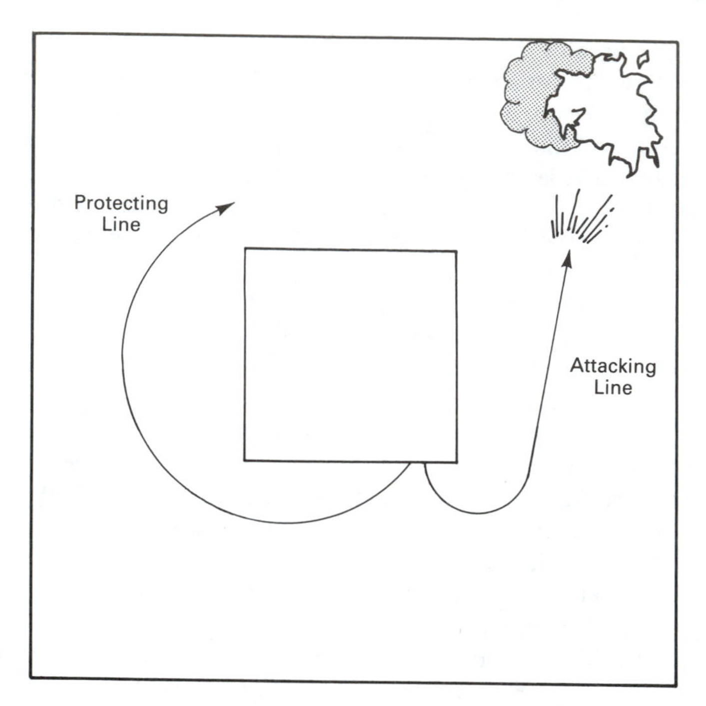

Figure 8.8 Provide a safety line as protection against a wraparound

Figure 8.9 Be on the lookout for the fire spreading overhead

from the time of receipt of the alarm until water was first applied to the fire. The time frame for getting water on the fire was not considered excessive for a fire in a high-rise building.

The rescue problem is considerably different at a high-rise fire. At most fires rescue consists of achieving a complete evacuation of the building. This is an impossibility in high-rise buildings. Some of these buildings house the population of a small city. The time factor, combined with the limited number of people available to perform rescue operations, may not permit complete evacuation. Consequently, other measures may have to be taken.

Water supply can be a problem. If the building is equipped with its own water supply system and an adequate pumping arrangement, the problem is somewhat limited. However, many of the taller buildings were constructed prior to the requirement for an adequate pumping system, which means that the responsibility for getting water to upper floors is completely that of the fire department.

Communication is always a problem at fires, but at high-rise fires the problem is compounded. Because of the type of construction it is almost impossible to get messages out of some buildings by the use of portable radios. Yet it is imperative that the incident commander be constantly informed of situations as they exist throughout the building. This means that the building's interior communication system or runners may have to be used. Some departments, however, have their own field telephones.

Having sufficient personnel to fight the fire is an acute problem. In fact, the total strength of most departments is not sufficient to cope adequately with a high-rise fire. It takes many times more people to fight a fire high in a building than it would if the fire was on the ground floor of a one-story building. Unfortunately, only a few of the larger cities have sufficient personnel and equipment necessary to do an effective job. For example, in 1988 the Los Angeles Fire Department used 64 fire companies and 383 firefighters on the First Interstate Bank Fire. Most departments will be hard pressed if a fire of any magnitude is encountered in their city. By necessity, smaller cities having high-rise buildings will have to depend on mutual aid for assistance in coping with these fires. It is extremely important that plans be made and training sessions held, as success depends on everyone at the fire being familiar with the tactics involved. Failure to do so will subject firefighters and occupants to an undue risk and most likely result in a predictable disaster.

Initial Response Responsibilities

The following is offered as a guideline for fighting fires in high-rise buildings. This guideline will undoubtedly have to be modified to fit the personnel and equipment capabilities and individual problems common to any particular city. The guidelines have been developed around the concept that any fire in a high-rise building could develop into a major problem.

The first alarm assignment. The objective of the first alarm assignment is to locate and extinguish the fire, if possible, while establishing the basis for operations if the fire develops into a multiple-alarm situation. This preparation includes establishing a fire attack team, a command post, a base station, and a staging area, and maintaining lobby control. Each of these terms will be defined later in the chapter.

The minimum response on the first alarm should be three engine companies, two truck companies, and a chief officer. Each of the companies should be adequately manned.

Duties of the first-in company officer. The responsibilities of the first arriving officer are to size-up the situation from outside the building, give a report of the conditions to the dispatch center, take command of the fire, and then lead the company (including the apparatus operator) into the building to locate and extinguish the fire. The first arriving officer's company will be known as the fire attack team.

Prior to entering the building the officer should determine that the crew has secured the proper equipment. They will need breathing apparatus, a portable radio, forcible entry tools or a rotary saw, and at least one high-rise hose pack. If possible, they should take along additional equipment such as a portable spotlight, extra air bottles, and a portable extinguisher.

Entrance to the building should be made through the lobby. Once inside, the officer should obtain information from the security personnel or building management as to the location and nature of the emergency. The officer should determine whether or not the building is equipped with an annunciator panel or a control room and obtain the phone number in the lobby. If an annunciator panel is available, it should be checked. The annunciator should indicate on what floor or floors the problem exists and whether the annunciator was tripped by a manual pull station, a smoke detector, a heat detector, or a water flow. Tripping of more than one device usually indicates that a problem exists.

Note: The following procedure for the use of elevators and stairways is based on the system used in the city of Los Angeles. Many fire departments in other parts of the country make greater use of elevators.

If there is a fire department lockbox in the lobby, the officer should take out one set of keys and a copy of the building inventory sheet, if available, and leave the remaining contents for the lobby control officer. The first arriving officer is then ready to determine the best and safest means of getting to the fire. The elevators should not be used until it has factually been established that the entire elevator shaft is not threatened by fire. An exception can be made if the building is equipped with a split bank of elevators in which the highest floor served is a minimum of five floors below the reported fire floor.

Finding the stairways is not always an easy task. The entrance to some blends in with the walls, which effectively camouflages the stairways. The doors to the stairways can normally be distinguished from other types of doors by their double-door width and key cylinder lock.

It should be remembered that all stairways do not serve every floor and extend through the roof. For example, scissor-type stairways may serve only every other floor. Some stairways extend only part way up the building, making it necessary to exit and find another stairway to continue the ascent.

Some stairways are equipped with a numbering system inside the stairway, which indicates which floors will be available and if the stairway extends to the roof. This information is invaluable when determining which stairway to use.

Once the determination has been made as to the best stairway to use, the company is ready to start climbing. Before doing so, the company officer should relay the stairway identification to incoming companies. If it is impossible to contact incoming units via radio, the officer should send the apparatus operator outside and have him or her make the contact.

If possible, the company officer should keep incoming units informed of what is found during his or her ascent. It is good practice to make an appraisal of conditions every four or five floors by opening the stairwell doors and checking inside. The ascent should be paced so that company members will not be completely exhausted upon reaching the fire floor. If so, they will be in no condition to attack the fire.

The second floor below the reported fire floor should be inspected to determine if it will be suitable for the staging area. The company officer should relay the survey results to the incident commander.

Upon reaching the fire floor, the company officer should determine what is burning and the potential for vertical and horizontal extension. The individual should determine if any occupants are endangered, and the best route for resources to move from the staging area to the fire floor. This information should be relayed to the incident commander. The company officer then assumes the duties of a division commander. A division is a geographical area, usually one floor in a high-rise incident. The division commander is identified by the floor on which he or she is in command. Upon assuming these duties, the division commander will attack the fire. If the fire cannot be contained with the available resources, the company should make every effort to protect the vertical openings and contain the fire until help arrives.

Duties of the second-in company officer. Upon arrival, the second-in company officer will enter the lobby and assume the duties of the lobby control officer. This officer is responsible for controlling all vertical access routes (including the elevators), controlling the air handling system, and coordinating the movement of personnel and equipment between the base and the staging area.

This officer should designate which stairways will be used for particular purposes and post personnel at each location to control entry and to direct civilians exiting from the building. He or she should also call all elevators to the lobby by the use of the emergency service control. The elevators should be secured there until it is determined that they are safe to use.

Elevators are the most effective means of transporting personnel and equipment aloft; however, if they are used improperly in a fire situation, personnel can be exposed to serious risk. They should not be used until it has definitely been established that the shaft and terminum are not threatened by fire and that there will be no disruption of the electrical power.

Duties of other first alarm officers. One of the other first alarm companies should be used to set up the staging area. The *staging area* is the assembly point where a reserve of personnel and equipment are awaiting assignment within the building. The minimum amount of reserve personnel and equipment to be kept at the staging area will be established by the incident commander. It is the responsibility of the staging officer to request additional resources whenever those in the staging area fall below the minimum.

The staging area is normally established two floors below the fire floor. The third floor below the fire will generally serve as an alternative if the second floor is not suitable because of a large collection of machinery, stock, or other material.

The base should also be established by one of the first alarm companies. The *base* is the reporting point for incoming companies and serves as the collection point for personnel and equipment pending transfer to the staging area. It should be located a minimum of 200 feet from the fire building. This is generally a sufficient distance to avoid the congestion around the building and is a safety factor against falling glass. Normally two companies should be kept at the base for each company held in the staging area.

Incoming company commanders should report to the base officer, who will give them instructions as to where and how to park their apparatus. Apparatus are normally parked diagonally so that they can be moved independently. Company officers should keep their personnel together at their apparatus while awaiting assignment.

The first arriving chief officer will assume the duties of incident commander upon arrival at the emergency. The chief should establish the location of the command post at least 200 feet from the fire building. The location of the command post should be relayed to the dispatch center.

The basis for controlling a large fire will have been met with the establishment of the command post, the fire attack team, the lobby control, the staging area, and the base. Any expansion or augmentation of this organization will be made by the Incident Commander after receiving information from the fire attack commander that a need exists.

Access Operations

Gaining entrance into a high-rise building is generally not a problem. The building is open during business hours and security personnel will normally meet the first arriving unit during nonbusiness hours. However, there are times when the building is locked and no one is on the premises to meet incoming units. In these instances not only will forcible entry probably have to be made into the building but also into each of the different occupancies within the building since they will most likely be individually locked.

Before forcing entry, make sure that a fire department exterior lockbox is not available. If one *is* available, it is usually in a location that requires a ladder to reach it.

If forcible entry is necessary, it should be accomplished in a manner that will result in the least amount of damage. At times the task can be accomplished by the use of two pry tools working together to force open entry doors. It may also be possible to cut the bolt between the doors or between the door and the floor with the use of the rotary saw.

If it is not possible to force a door, it may be necessary to break glass to gain entrance. The entrance doors are normally made of extremely thick glass, which makes them difficult to break. In addition, they are very expensive to replace. An attempt to break the glass in these doors should be made only as a last resort. A better choice is the glass panels adjacent to the entry doors. These panels are generally made of thinner glass and are easier to break.

Consideration for gaining entrance should not be limited to the front entrance. At times the building has a basement with a stairway leading from the outside into this area. Gaining entrance at this point is normally quicker and easier than through the front; however, the initial action within the building commences in the lobby, which makes it better if entry can be made directly into this area.

On some high-rise buildings it may be possible to raise a ladder and gain entrance through an upper level. This is not possible, however, in many of the newer buildings because the windows are unopenable.

If fire is showing upon arrival of the first unit, priority should be given to gaining entrance as quickly as possible.

Rescue Operations

Rescue operations in high-rise fires are considerably different than in other types of fires. For example, a fire in a dwelling generally places all occupants in immediate danger. This is particularly true when the fire occurs at night. This is generally not true at high-rise fires. Only a few, if any, of the occupants are in immediate danger when the department arrives.

Occupants of high-rise buildings can be placed in danger in the event of a fire by three primary methods. Those in the vicinity can be directly exposed to the *flame*. This normally occurs in residential buildings where the fire is confined to a single unit. Others in adjacent units, however, may be exposed if the fire has a pathway for extension.

The most likely and by far the greatest threat to occupants from the fire is exposure to *smoke* or the other products of combustion. Smoke may travel up elevator shafts, stairwells, and other unprotected vertical openings. It may also extend through air handling systems in older buildings that lack automatic shutdown devices. The newer buildings are designed with such devices as compartmentalization, pressured stairwells, and pressured elevator vestibules to restrict smoke travel, but these measures are not always successful.

The third threat is that of *panic*. Panic can occur when occupants know or suspect that a fire exists. The fire does not have to be in the immediate vicinity to cause irrational behavior.

As with fires in most other buildings, the best means of saving lives and facilitating rescue operations is an aggressive attack on the fire. This, together with proper ventilation methods, provides the highest degree of safety to occupants. However, it does not eliminate the need to make an adequate search if the fire is of any size whatsoever.

Search procedures. People can be found in high-rise buildings at all times of the night and day. It is important during fire operations that all occupants be accounted for. A complete search of a high-rise building is a long and tedious task and normally does not have to be made except when the fire is extensive. It is time consuming and requires a considerable number of personnel to do the job effectively. However, a limited search is normally required if the fire is of any size whatsoever. Search procedures should be organized, directed, and coordinated by a responsible officer.

The search should be organized on a priority basis. First priority should be given to the immediate area on the fire floor. Second priority should be given to the floor above the fire. The search should then continue, floor by floor, both above and below the fire until a complete search of the area deemed necessary has been made.

Search efforts should be conducted through the use of search teams, each headed by a responsible officer. The use of companies as search teams is ideal if a sufficient number of companies are available. Each search team should be provided with keys to gain access from the stairwell to individual floors and also room keys if such are available.

It is imperative that a strict accounting and documentation system be developed to ensure that every room on every floor is searched and that search efforts are not duplicated. (See Figure 8.10.) The method used by one of the larger departments is worthy of consideration. As the search team enters the floor from the door leading from the stairway, a large diagonal line is made with a piece of chalk on the door. The search team's identification number is placed below the diagonal line. A similar diagonal mark without the identification number is placed on the door lead-

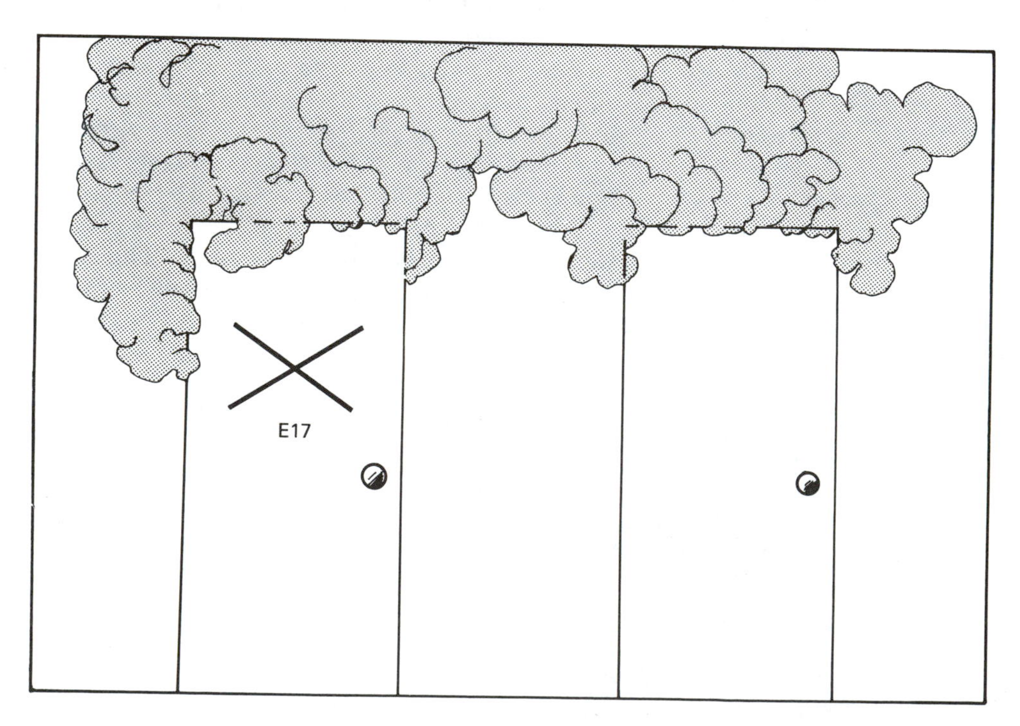

Figure 8.10 Avoid duplication of effort

ing to an office, dwelling unit, and so on. When a complete search has been made of the unit, the searching member or members close the door and place another diagonal mark on the door. This makes the original diagonal mark into an "X." Upon completing the search of all rooms on the floor, the search team makes the "X" on the stairwell door and moves on to the next floor that needs searching.

Relocation of occupants. Total evacuation of a high-rise structure during firefighting operations is neither practical nor feasible. The primary effort should be devoted to moving those people who are threatened by the fire to a place of safety. This normally requires the movement of those on the fire floor and two floors above and below the fire floor. Occupants of other than these five floors should be instructed to remain where they are, unless of course there is a chance that they may be placed in jeopardy. Those on the fire floor and the two floors below the fire floor should normally be moved to the floor below the staging area. Those on the two floors above the fire floor should normally be moved to the third floor above the fire floor. In some cases, however, it may be necessary to move all those above the fire to the roof.

Providing information and instructions to those in public areas within the building is simplified if the building is equipped with a well-designed communication system. These systems allow a fire department member to talk to occupants on individual floors or given areas of the building as the situation deems necessary. The system is normally controlled from the fire control room. The individual chosen to send the messages should be one who is capable of projecting confidence. In cities where a number of the occupants speak a language other than English, it is wise to repeat the instructions in the appropriate language.

Controlling the evacuees. Regardless of the department's desire to keep occupants where they are, some will leave. There is no harm to this as long as they do not hamper firefighting operations. It is best to designate which stairway should be used by occupants for evacuation and then maintain control of this pathway. The important thing is to keep the traffic moving and not to let bottlenecks form.

Once occupants have left the building, they will have a tendency to stop and talk. This not only puts them in jeopardy from the possibility of falling glass but can create a logjam, which could extend back into the building and slow down the orderly evacuation of other occupants. To prevent this from happening, a corridor should be established that will lead occupants to an assembly area away from the building. It is a good idea to use fire line tape or rope to designate the corridor. Police officers and/or security personnel can be used to keep the evacuees moving.

All evacuees should be logged in once they reach the assembly area. The information should include where they work or live within the building. If the individuals have no medical or other problem relating to the emergency, they should be free to leave the assembly area. They should be instructed upon leaving that they are not to return to the building. Whether or not they leave the scene is between them and their individual employers.

Firefighting Tactics

The basic principle of fighting a fire in a high-rise building is no different than that of fighting a fire in any other type of building. The time-honored sequence of locate, confine, and extinguish prevails.

If ascent has been made by way of the stairwell, the basic operations will begin when the reported fire floor is reached. As when entering any other potential fire area, the door should be felt and a check made for smoke. The door can be opened if indications are that it is safe to do so. If no fire can be found, the floor both above and below should be checked, as a malfunction may have caused the alarm system to indicate the wrong floor.

If fire is suspected on the reported fire floor, the door should not be opened until a charged line is in place. The line may be connected to the outlet on the fire floor or the floor above or below the fire floor, depending on the circumstances encountered. A 2½-inch line should be used, as it is always easier to reduce the line than it is to increase it. The line should be equipped with a shutoff arrangement that permits the removing of the tip if it becomes necessary to extend the line.

It is good practice before opening the door to advance the hose line to ensure that the stairwell is not being used by occupants of the building for egress. The door cannot be closed once it is opened and the line is taken inside. This means that smoke and other products of combustion can escape from the fire area and quickly make the stairwell unsuitable for occupants. If occupants are using the stairwell, it may be necessary that entry be delayed until all evacuees have passed the fire floor.

If the door is hot to the hand, it may indicate that a potential backdraft condition exists inside. This condition could be caused by the shutting down of the heating and ventilation system, which might result in an oxygen deficiency in the fire area. In case of a hot door, the door should be opened with caution. If air rushes in when the door is opened, the door should be immediately closed and entry not be attempted until the area is ventilated.

Venting the fire floor of a high-rise building is not easy. One method is to break the windows with a ball or chain or by the use of a heavy object such as an axe attached to a drop line. The incident commander should be notified before any glass is broken so that the area below can be cleared.

The windows can be broken from two floors above the fire floor. However, it may also be necessary to break or remove glass on this floor to provide a place from which to work if the building is equipped with unopenable windows. Charged lines should be in position on the floor above the fire floor and the floor from which the windows will be broken prior to breaking any glass. The objective of these lines is to prevent the fire from leaping out the broken windows on the fire floor and extending into upper floors. The fire attack team should be ready to move in with charged lines once the windows are broken.

Breaking glass on an upper floor is a risky procedure. It is possible that the vacuum or winds could take a firefighter out a window. It is good practice to secure the firefighter who is breaking the glass with a safety rope prior to the glass being broken.

The potential for a flashover is always present in a high-rise fire. One of the contributors to this potential is the extensive use of plastics and other synthetics in the furnishings and decorations. Firefighters entering the fire area should constantly be alert for this possibility. Some of the indicators of a possible flashover are a freely burning fire and an accumulation of extensive heat and/or smoke at ceiling level. The smoke may sometimes remain at the ceiling level or it may bank down. The company officer should consider that the potential is extreme if firefighters are driven to the floor by extensive heat from above. It is acceptable at this point to project a spray stream across the ceiling to reduce the built-up heat.

Elevators are the quickest method of moving personnel and equipment to upper floors if it has been determined that they are safe to use. Normally their use is restricted to later-arriving companies, as the decision for their use is based on information relayed to the incident commander from the initial attack team. If an elevator is used to take an attack team aloft, the team should unload two floors below the fire floor and proceed to the fire floor by way of the stairwell as a precautionary measure against the fire meeting them when the elevator doors open.

Movement of Personnel and Equipment

Movement of personnel and equipment to the fire is accomplished through a relay operation. The relay points are the staging area, the base, and the lobby control.

The staging area. The staging area is the collection point for the equipment and personnel that will be used on the fire. It is normally located two floors below the fire floor. A determination as to the minimum amount of equipment and the minimum number of personnel to be held in reserve is made early during the emergency. Normally the equipment supply list will include air bottles, hose, fittings, breathing apparatus, forcible entry tools, salvage equipment, pike poles, ladders, and a resuscitator. The supply is replenished as it is dispatched to the fire. Empty bottles that have been used on the fire are also stored in this area.

The used equipment is separated from that which is available for dispatch. The available equipment is separated by type with signs taped to the wall indicating individual storage areas.

The staging area is also the collection point for fresh personnel and those who have been relieved from the fire. These two groups should be separated so that they can be readily identified. Requests to replenish supplies or personnel are made to the base.

The base. The base is the collection point for incoming companies and equipment. It is located at least 200 feet from the fire building. Normally two companies are kept in the base for each one held in reserve in the staging area. Equipment or personnel requested from the staging officer is sent to the lobby control officer.

Lobby control. Lobby control will deliver the equipment to the stairwell entrance at ground level where it will be received by the stairwell support group. This group is responsible for moving the equipment from the ground level to the staging area floor hallway. It consists of personnel stationed on every other floor from the ground floor to the staging area, with an officer supervising every four or five people. One member picks up the equipment at the ground floor entrance to the stairwell and carries it to the third-floor landing. He or she then returns to the ground level to pick up a second load. Another member picks up the equipment at the third floor and carries it to the fifth. This process continues until the equipment reaches the staging area hallway. At this point it is picked up by a member from the staging area and carried to the proper storage area.

Maintaining records. It is important that records be kept at each stage of the relay. The base officer should log in all companies and equipment that report to the base. All entries should include time. He or she will also log in all requests for equipment or personnel from the staging officer and log out what was sent.

This same logging in and out procedure continues with the lobby control officer.

The staging officer logs all requests made for equipment and personnel. In addition, he or she logs in all equipment or personnel received from the lobby control officer and all sent to or received from the attack forces.

Air Support

Those departments having helicopters normally include this equipment in the plans for fighting fires in high-rise buildings. Helicopters can be used effectively for reconnaissance, delivering equipment, and rescue. At times, the use of helicopters is the only effective means of rescuing people who have been trapped above the fire floor. If endangered, these people will generally move to the roof. From here they can be rescued by helicopter, providing the heat and smoke do not make flight operations impractical.

Sometimes it is easier to move equipment from the roof to the staging area than it is to move it from the ground floor to the staging area. This is particularly true if the fire is high in the building and the building is equipped with smoke towers that permit a downward movement of supplies. In these instances it may be advisable to move a stairway support group to the roof to facilitate the transportation of equipment.

Helicopter operations are placed under the direction of an air operations director who reports to the operations chief. The helispot should be located at least a half-mile from the incident site. Operations at this location will minimize the confusion caused by the noise and rotor downwash from the aircraft.

Multiple-Alarm Fires

Adequate establishment of the basic command system by the first alarm companies provides the pathway for smoothly moving into the organization required for coping with a major fire. The Los Angeles Fire Department's Incident Command System for fighting high-rise fires provides for four officers reporting to the incident commander.

- The operations chief is responsible for all firefighting operations. The staging officer, the air operations director, and all division commanders are under this individual's command.
- The logistic chief is responsible for logistic support. The lobby control officer, the base commander, the stairwell support unit, and communications are under this individual's command.
- The planning chief is responsible for documentation of the incident and keeping the incident commander apprised of the status of the various phases of the operation. This individual assists the incident commander in planning strategy.
- The command staff chief manages the staff functions of safety, liaison, and information. This individual has a staff to assist in each of these areas.

Movement from the basic system established by the first alarm assignment to the full-blown incident command organization is accomplished on a priority basis. The first arriving battalion chief assumes the duties of incident commander. The second arriving battalion chief is assigned the duties of operations section chief. The first arriving assistant chief will relieve the incident commander (first arriving battalion chief). This chief will then become the planning chief.

Any additional command level personnel will be assigned at the discretion of the incident commander.

REVIEW QUESTIONS

1. What are two of the many factors that contribute to the differences between fires in buildings?
2. How much water is required to extinguish the majority of all fires found in buildings?
3. What is the value of interior standpipe lines to firefighting personnel?
4. What is the value of exterior dry standpipe systems to firefighters?
5. Which lines laid into sprinkler systems are particularly effective?

ATTIC FIRES

1. What size lines should normally be used to attack attic fires?
2. What type nozzle is preferred?
3. What is the guideline as to the amount of water to use to extinguish the fire?
4. When should salvage operations be commenced?
5. At what size fire should the roof normally be opened?
6. What is an effective way of extinguishing the fire if it is in a large finished attic with a gable roof and access from below is difficult?

BASEMENT FIRES

1. How does the record of basement fires getting out of control compare with the record of other fires getting out of control?
2. How fast do basement fires burn once ignition takes place?
3. What is one of the first things that should be done upon arrival at a basement fire?
4. Where should the first lines be placed?
5. What factor has the biggest influence on determining the best method of attack?
6. What procedure should be used if there are openings into the basement from both ends?
7. What procedure should be used if it becomes necessary to open the floor for ventilation?
8. What may be necessary if the basement is well involved with fire?
9. What is a good practice to use when cellar devices are discharging water into the fire area?
10. When should the use of cellar appliances be discontinued?
11. What is the value of high expansion foam at cellar fires?
12. What can be used as a last resort to extinguish the fire?

CHIMNEY FIRES

1. How serious a problem do fires in chimneys and flues normally present?
2. What is the best method to use if the fire is contained within the chimney?
3. If it is necessary to use water, what practice should be observed?
4. How can the chimney be cleaned out if it is stopped up?
5. What should be done before this takes place?
6. What would smoke in the attic generally indicate if the fire was in the chimney?

PARTITION FIRES

1. What is the chance of a rekindle from a partition fire?
2. In what type of structures is the chance for a partition fire most prevalent?
3. What is the best visual indication of a partition fire?
4. What is the best method of detecting fires in partitions?
5. If it is necessary to open a wall to check for a possible partition fire, where should it first be opened?
6. How should the stream be directed once the wall is opened?

DWELLING FIRES

1. What is the potential dollar loss in some dwelling fires?
2. What is the primary problem in dwelling fires?
3. Which is most common: single-room fires or well-involved dwellings?
4. Where should the first attack line be brought in on a single room fire?
5. What size lines are normally adequate for well-involved, one-story dwellings of average size?
6. What firefighting tactics should normally be used on this size fire?

7. What is good practice regarding the first attack line for fires in larger one-story dwellings?
8. Which dwelling fires are most feared by firefighters?
9. Where are the most threatening fires found in multiple-story dwelling fires?
10. To what location should the first lines normally be advanced in multiple-story dwelling fires?
11. What is a good practice to use if large volumes of fire are visible upon arrival at a multiple-story dwelling fire?

GARAGES

1. What two construction changes have increased the problem in garage fires?
2. How is the attack on a garage fire generally made?
3. What is a good procedure to use when the large garage door is opened?
4. What procedure should be used to protect a one-story dwelling when the fire is in the attached garage?
5. What procedure should be used to protect a two-story residence if the fire is in the garage, which is attached at the first floor and under the second floor?
6. Why should extreme care be used when making the attack on the main body of fire in a garage?
7. What would be the best indication that magnesium is involved in a garage fire?

MERCANTILE FIRES

1. What are the construction features of a typical mercantile store?
2. Where do most fires start in mercantile occupancies?
3. What is the potential for a fast-moving fire?
4. What is the value of salvage and ventilation operations?

INDUSTRIAL FIRES

1. Should a larger number of serious fires be expected in industrial plants or in mercantile establishments?
2. What type of streams are many times necessary for use in the initial attack?
3. What type of firefighting tactics are generally used?
4. What type of streams are particularly effective for use in protecting exposures of large quantities of flammable liquids?
5. What is the hazard of mingling acids and cyanides?
6. How should fires in areas where molten metals are found be extinguished?
7. How should fires in dust-collecting bins be extinguished?
8. What precautionary measure should be taken if vats of hot salt baths are present in the fire area?
9. What precautionary measure should be taken when fires occur in dip tanks or large vats of combustible paints?
10. What type streams are generally required for fires in lumberyards or in the lumber storage area of industrial plants?

PLACES OF PUBLIC ASSEMBLY

1. What was the loss of life in the following fires: Iroquois Theater, Cocoanut Grove, and Beverly Hills Supper Club?
2. What was the big contributor to the loss of life in each of these fires?
3. What precautionary measure can the fire department take to help eliminate panic?
4. What is one of the first actions that should take place by fire department personnel after arriving on the scene?
5. For what purpose should the first lines be used?
6. What action can be taken to help eliminate panic in the event that the building is thrown into darkness?
7. Where is the most likely place for a fire to start in a theater?
8. What action should the officer-in-charge take when responding to a fire backstage in a theater?

TAXPAYER FIRES

1. What is the origin of the term *taxpayer* fire?
2. What are the construction features of a taxpayer?
3. What is the primary problem in this type fire?
4. What action should be taken to help alleviate this problem once the location of the fire is determined?
5. What is the minimum size hole to cut in the roof?
6. What size lines are normally used to stop the spread in the attic?
7. What is the primary difference between the firefighting problems in a one-story taxpayer and a two-story taxpayer?

CHURCH FIRES

1. What should be anticipated if the fire is of any size at all?
2. What are some of the construction problems?
3. Where can some of the concealed spaces be found?
4. Where do a large number of the fires originate?
5. What is the chance of a fire spreading from this location?
6. What action should take place at the fire if the church has an unprotected passageway extending from the church to the rectory?
7. What size lines should be used if the fire has gained headway?
8. What can be done to extinguish the fire if it has entered the hanging ceiling or enclosed attic area?
9. What precautionary measure should be taken regarding the placement of apparatus that will be used to provide elevated heavy streams?
10. What thought should be kept in mind regarding personnel working inside if the fire is of any size whatsoever?

FIRES IN BUILDINGS UNDER CONSTRUCTION

1. Fires in what types of buildings under construction appear to present the most troubles for firefighters?
2. At what point during the construction stage are these buildings most hazardous?
3. When is the hazard from a fire that occurs during this stage of construction the greatest?
4. What are the general problems in this type of fire?
5. In reality, what is the only method of effectively fighting a fire in these types of complexes?
6. What should be included in ordinances controlling construction in these complexes?
7. What is the primary problem once a fire starts in these complexes?
8. How can exposures be protected?

HIGH-RISE FIRES

1. What is a high-rise building?
2. What three factors were merged in the development of high-rise firefighting tactics?
3. What should the water supply system for a high-rise building contain?
4. How can the local fire warning system in a high-rise building be actuated?
5. What will the alarm mode do?
6. What capabilities are in the one-way communication system?
7. What does the building control station house?
8. What is found in the fire department lockbox?
9. What are some advantages and disadvantages of the type of construction found in high-rise buildings?
10. What are the two basic types of high-rise buildings?
11. What is the principle of the center-core design concept?
12. What is the problem with the construction features of the ceiling?
13. What is one of the primary problems with fighting fires in high-rise buildings?
14. What are some of the other problems?
15. How should smaller cities try to cope with the high-rise problem?
16. What should be the objective of the first alarm assignment?
17. What should be the minimum response on the first alarm?
18. What equipment should the first company to arrive take into the building with them?
19. Where should entry be made into the building?
20. What are some of the factors that should be determined in the lobby?
21. What is the restriction on the use of the elevator?
22. Which floor should be surveyed for possible use as the staging area?
23. What should the company officer determine when reaching the fire floor?
24. What is a division?
25. What are the duties of the second-in company officer?
26. Who is responsible for setting up the staging area?
27. Where is the staging area normally set up?
28. Where should the base be set up?

29. Where should incoming company commanders report?
30. What does the first arriving chief officer do?
31. Where should the command post be set up?
32. What are some of the means of gaining entrance if forcible entry is required?
33. What are the three primary methods by which occupants of the building can be placed in danger from the fire?
34. What is the most likely and by far the greatest threat to occupants from the fire?
35. What is generally the best means of saving lives and facilitating rescue operations?
36. Where should first priority be given when making a search?
37. Where should second priority be given?
38. What should be provided to search teams?
39. What is one method to eliminate duplication of search efforts?
40. Normally, which people should be moved to a place of safety?
41. From what location is the one-way communication system operated?
42. What method should be established for controlling evacuees?
43. What is the basic difference between fighting a fire in a high-rise building and that in a different type building?
44. What should be done if no fire is found on the fire floor?
45. What size line should be used for the initial attack?
46. What type arrangement should be on the tip?
47. What precautionary measure should be taken regarding the stairwell before the line is advanced?
48. What should be done if the door is hot to the hand?
49. What should be done if air rushes in when the door is opened?
50. What is one method of venting the fire floor?
51. What should be done prior to doing this?
52. What are the contributors to a potential flashover?
53. What action might be taken if firefighters are driven to the floor by extensive heat from above?
54. Where should the crew get off the elevator if it is used for taking a fire attack crew aloft?
55. What are the relay points for movement of personnel and equipment to the fire?
56. What is the collection point for the equipment and personnel that will be used on the fire?
57. What equipment is normally held in reserve in the staging area?
58. What is the collection point for incoming companies and equipment?
59. What does the stairwell support group do?
60. How are they spaced when equipment is being moved?
61. Who maintains records of incoming and outgoing equipment and personnel?
62. For what purpose can helicopters be used?
63. What four officers report to the incident commander in the Los Angeles Incident Command System?
64. What is the operations chief responsible for in this system?

65. What is the logistic chief responsible for in this system?
66. What is the planning chief responsible for in this system?
67. What are the duties of the first-arriving battalion chief in this system?
68. Where is the second-arriving battalion chief assigned in this system?
69. What are the duties of the first-arriving assistant chief in this system?
70. What position does the first-arriving battalion chief take when relieved as incident commander by the assistant chief?

CHAPTER 9

Fires in Mobile Equipment

OBJECTIVE

The objective of this chapter is to introduce the reader to firefighting tactics as they apply to mobile equipment. Car fires, truck fires, aircraft fires, small boat fires, and ship fires are explored. Review questions are provided at the end of the chapter to assist the reader in determining his or her level of comprehension of the chapter contents.

Although the basic firefighting tactics are applicable to fighting fires in mobile equipment, there are some peculiarities that apply to individual units that are worthy of consideration. One factor that applies to all is that of surprise. Fires occur in places where a few minutes before no problems existed. In most cases the fire department will be faced with a situation where it is not exactly sure what is burning or how much of the material is involved. At times firefighters will be able to determine what materials are being transported prior to attacking the fire, but in most instances the attack will be made and the determination as to what was involved has to wait until after extinguishment has been completed. However, some basic tactics can be developed that will not eliminate all risks but will reduce them to a minimum.

In the development of these tactics, it should be remembered that any fire of hazardous materials or where hazardous materials are involved requires a cautious and fully protected approach. In some cases it is better to evacuate and let the fire burn. If a hazardous material team is available, it should always be requested in these situations.

CAR FIRES

At first glance, the extinguishment of a fire in a car appears to be a simple task. It should be kept in mind that car fires account for a large number of fatalities to occupants, and a number of firefighters have been severely injured as a result of

Photo 9.1 Cars can become well involved with fire in a short period of time. (Courtesy of Chris Mickal, New Orleans Fire Department Photo Unit)

not following good safety precautions. The primary problem to firefighters is the chance for a rapid spread of the fire due to a spill of gasoline, a rupture of the fuel tank, or a rupture of hydraulic fluid lines. The rupture of hollow drives lines, gas-filled shock absorbers, shock-absorbing bumpers, and tires, as well as the burning of toxic plastics are also major concerns. Adequate steps must always be taken to guard against these potentials.

One of the first precautionary measures to take is when arriving at the fire. Where the fire apparatus is parked will depend on a number of circumstances. It is important to park the apparatus in such a manner as to protect the firefighters from moving traffic. If the burning car is on an incline, it is best to park the apparatus uphill at least 150 feet from the car. This will prevent the car from rolling into the apparatus in the event the brakes give way. Regardless, it is a good idea to place a chock block under the wheels of the car to prevent its movement during the extinguishment phase. Thought should also be given to protecting the firefighters in the event of a fuel tank rupture. It is good practice to spot the fire apparatus to the front of the car, even if it means passing the vehicle on the way in. It is particularly important to follow this procedure if the fire is in the rear where the gasoline tank could become involved. Of course, if the fuel tank is in the front of the car, the apparatus should be parked to the rear.

At most car fires, the responding company will find smoke issuing from under the hood or from inside the body of the car. If the smoke is coming from under the hood, it is necessary to raise it to get at the fire.

If there is access to the interior of the car, the hood latch should be pulled to unlock it. This latch is normally found under the dash on the left-hand side. Care should be taken, however, when opening the car door to release the hood. It is possible that smoke and heat have penetrated the fire wall and built up inside the car. There have been cases where a backdraft occurred when the door to a car or van has been opened when all signs indicated that the fire had been confined to the engine area. It will also be necessary to release a second latch, which is generally found directly under the hood. Gloves must be used for this purpose because the metal will undoubtedly be hot. Of course, the firefighters should already have gloves on, as each of them should be equipped with full protective clothing including breathing apparatus.

If the metal is too hot to attempt to release this secondary latch, a pike pole or similar tool can be used to pry open the hood. The fire can generally be knocked down with a small spray stream from a water extinguisher or a small line once the hood is opened.

Some officers prefer to knock the fire down before completely opening the hood. This can be done by using a CO_2, dry chemical, or halon extinguisher through a crack or by applying a fog stream from ground level up into the engine area. Regardless of which method is used, there is a likelihood of a reflash once the hood is opened if a followup is not made with water to cool down the engine. If the fire under the hood is severe, the knockdown by one of these extinguishers followed by water is usually more effective and less hazardous than opening the hood and using water.

A fire under the hood will generally damage part of the electrical system. It is good practice to disconnect the battery any time the fire is in any part of the engine or in any electric circuit. This can be done quickly with the use of a screwdriver and a pair of pliers.

Fires inside the car generally involve the upholstery. These fires can normally be knocked down by the water from an extinguisher or a small line. There is, however, a possibility that the fire may have extended into the trunk if it is severe and includes the back seat. Under these conditions it is necessary to open the trunk. The trunk lock can sometimes be released from inside the car. If not, and if the keys are not available, the trunk will have to be pried open and checked for fire extension. A charged line should be in position when the trunk is opened, as it is impossible to determine what might be involved inside, and there is always the possibility that backdraft conditions prevail.

Backdrafts in cars and vans are not uncommon. Care should always be taken to avoid this possibility when opening up any portion of the vehicle.

Sometimes the fire will be found around the gasoline tank. If there is fire at the mouth of the tank, do not remove the cap until the tank has been adequately cooled.

Occasionally fully involved cars are encountered. These should be approached with caution. It is good practice to open up on these fires with a straight stream from the maximum distance. The stream should be first directed at the gasoline tank to keep it from rupturing or venting under pressure, and then the tires to prevent a similar action. The stream can be changed to a spray stream as the car is approached. *Do not* make the initial attack from the rear of the vehicle. (See Figure 9.1.) Stay away from this area until assured that the tank has been cooled sufficiently to eliminate the threat of a rupture. Remember that some of the fuel tanks are made of plastic.

The life hazard in a car fire can be severe. A person trapped in a burning

Figure 9.1 Do not attack the car from the rear

vehicle is a frightening sight. The only logical means of rescue is normally an aggressive attack on the fire. The victim could be unconscious or physically trapped. If unconscious, it may be simply a matter of releasing the seat belt and removing him or her. Seat belts, however, sometimes present a problem. It is good practice for firefighters to carry a knife that is capable of cutting the seat belt if necessary.

Vehicle extrication procedures will have to be used to remove the occupants if they are physically trapped. Care should be taken during this operation to ensure that a reflash will not occur due to leaking gasoline. All potential sources of ignition should be removed, hot parts of the engine cooled, and a loaded line be kept available until all occupants have been safely removed.

TRUCK FIRES

Truck fires in or around the engine or inside the cab are handled the same as those found in cars. Rescue practices are also similar. The problem with truck fires is the cargo. Firefighters are generally not aware of what is being carried. Signs indicating the presence of hazardous materials may be displayed on the truck but it is possible that these hazards are present without the luxury of warning signs.

Fires in trucks should always be suspected as carrying hazardous materials and approached as such. Full protective gear with breathing apparatus is a must. It is good practice to attack a fire in the cargo hold of a truck from the upwind direction from a maximum distance, using a straight stream for the initial attack. Thoroughly cool the gasoline tank and tires as advancement is made. Spray streams can be used as the fire is closely approached. Be on the alert for a possible backdraft as the cargo space is opened. The manifest of what is carried should be checked prior to overhaul. It can generally be found in the driver's compartment. Standard overhaul procedures should be modified in accordance with what is found in the manifest. The hazardous materials team should be requested, if available, if hazardous materials are indicated in the manifest.

In some cases, depending on the evidence, it is best to let the fire burn and evacuate to a safe distance. This should be done if there is any doubt as to the safety of personnel.

AIRCRAFT FIRES

Regardless of the size or location of the community or the size of the firefighting force, a fire department faces the possibility of having to fight a fire in an aircraft at any hour on any day. The majority of all aircraft crashes have occurred on or within five miles of an airport. However, aircraft have crashed in some of the most unlikely places. In fact, some of the most disastrous incidents have not happened within five miles of an airport. This section will cover some of the basic fires that

Photo 9.2 Aircraft crashes can occur anywhere, anytime. (Courtesy of Chicago Fire Department)

occur in aircraft on the ground, together with those resulting from crashes or forced landings.

Engine Fires

Fires in engines of aircraft on the ground are normally not too serious. The procedure for handling the fire will differ, depending on whether it is in a piston or a turbine (jet) engine.

Fires in piston engines. If the fire is confined within the engine nacelle, the pilot or maintenance crew should first try to extinguish it by using the aircraft's built-in extinguishing system. If this is not successful, the fire should be attacked using either carbon dioxide or halon. If deemed necessary, water spray or foam can be used to keep the adjacent parts of the aircraft structure cool.

Fires in jet engines. Fires in the combustion chamber of jet engines are best controlled if the engine can be kept turning over by the crew. Extreme caution should be observed, however, when working around a running jet engine. Personnel should never stand within 25 feet of the front of, to the side of, or directly to the rear of the engine intake ducts. The suction created by some engines is sufficient to draw in a 200-pound person. Personnel should also stand clear of the turbine plane of rotation area. In the event of engine disintegration, this area is the path of flying parts. The area to the rear of the exhaust outlet should also be avoided for a distance of at least 150 feet. Exhaust temperatures reach approximately 3,000° F. at the outlet.

Extinguishment of fires that are outside the combustion chamber but within the engine nacelle would first be attempted by using the aircraft's built-in extinguishing system by the pilot or maintenance crew. If this is unsuccessful, then carbon dioxide or halon should be tried; however, these extinguishers should not be used if magnesium or titanium parts are involved. In these cases it is best to allow the fire to burn itself out. Foam or water spray should be used to keep the nacelle and surrounding exposed parts of the aircraft cool while this is taking place.

Wheel Fires

Potential wheel fires should be approached with caution. The responding fire apparatus should be parked within effective firefighting distance from the aircraft but never to the side of the aircraft or in line with the wheel's axle. If the tire should explode, debris is generally thrown to the sides but not to the front or rear. Consequently, if it is necessary for firefighters to approach the aircraft, it should be done from a fore or aft direction. (See Figure 9.2.)

Smoke around the drums and tires of the wheels does not necessarily mean that the wheel is on fire. Overheating of brakes is not an unusual occurrence on aircraft. If the brakes are overheated, they should be allowed to cool by air only. On propeller-driven aircraft the cooling can be assisted if the crew will keep the

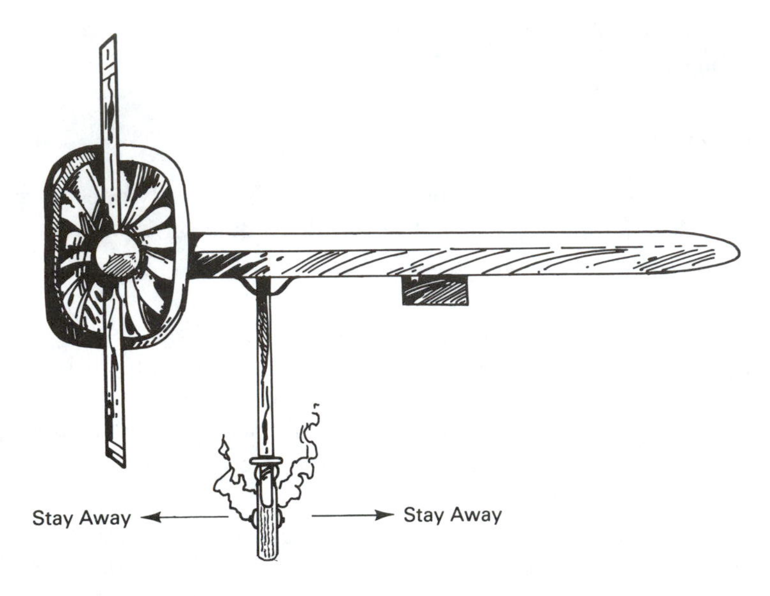

Figure 9.2 Remain clear of the danger area

engine running, which will direct fresh air onto the wheel. Firefighters should remain at a safe distance and not approach the aircraft while this is being done.

It is possible for firefighters or members of the ground crew to use a smoke ejector to help reduce the cooling time in the event the aircraft's engines cannot be used for cooling or a jet aircraft is involved. Do not use water for cooling, as the rapid reduction of heat in the wheels may result in an explosion. Carbon dioxide extinguishers are frequently used for cooling wheels.

An extinguisher can be used if there are actual flames in the wheel. A dry chemical extinguisher is recommended because it is less likely to chill the metal in the wheel parts. Water can be used if a dry chemical extinguisher is not available; however, it should be used with caution. Firefighters should protect themselves from a possible explosion by using the fire apparatus as a shield or by standing fore or aft of the wheel. The water should be applied in a fine spray and in short bursts of 5 to 10 seconds. At least 30 seconds should elapse between bursts. Water should be used only as long as visible flames are showing.

Interior Fires

Fires in the interior of an aircraft on the ground normally involve ordinary combustible material. It should be kept in mind that there is sufficient combustible material that gives off toxic gases on decomposition in the interior of most aircraft to warrant a standard procedure of wearing breathing apparatus to attack such fires, regardless of the size of the fire.

Fires originating when the aircraft is occupied are normally detected in the early stage, which generally results in the damage being restricted to a minimum.

However, the fire will probably smolder for some time before being discovered if ignition takes place when the aircraft is unoccupied.

Fires discovered in the early stages can normally be extinguished by the use of a hand extinguisher. Small lines, however, will have to be employed if the fire is beyond the control of hand extinguishers.

The interior of a larger aircraft is similar to a corridor in an ordinary structure that contains a considerable amount of combustible material. This type of arrangement always presents the potential for a backdraft or a flashover. Consequently, extreme care should be taken when entering the cabin and while working on a fire once inside. It goes without saying that firefighters should always be equipped with full protective equipment, including breathing apparatus. If the fire is small, it can generally be extinguished by making a direct attack with the use of small lines. On large commercial aircraft, the most stable point to initiate the attack is through the exits over the wing. If it is decided that it is best to enter the aircraft to fight the fire, the positive pressure ventilation method of firefighting might prove extremely effective. Open the top of the aircraft or a cabin door some distance from the entry point and have firefighters with blowers precede those with 1½- or 1¾-inch hose lines. This should help remove the atmosphere of smoke and products of combustion and make it easier for the nozzlemen to make entry.

The temperature inside the aircraft will probably be extreme if the fire is extensive. In these circumstances it might be best to use the indirect method of attack from outside the aircraft. Several lines should be used in a coordinated effort if such a selection is made. Where the lines should be positioned can be determined by observing the conditions through the various cabin windows, together with the blistering of the external paint. Care should be taken so that the lines are not positioned in such locations that will result in forcing the fire into the uninvolved portions of the cabin area, providing that there are any.

It is possible that the fire can extend into the hidden areas in the ceiling, walls, and overhead storage areas. These areas should be thoroughly checked once the fire is under control.

Aircraft Crashes

Aircraft range in size from those carrying 1 person to those transporting nearly 500 people. A crash involving an aircraft can occur in or near any city at any time. The magnitude of the problem facing firefighters will depend on a number of factors. Some of the primary factors are the seriousness of the crash, the size of the aircraft, the number of people aboard, and the personnel and equipment available to cope with the situation.

The primary objective in fighting a fire in an aircraft is the same as fighting one in any other occupancy—saving lives. However, the time frame is much more critical where aircraft fires are concerned. The primary reason for this is that flammable liquids are normally involved, which can result in an extremely intense fire that spreads rapidly.

The material contained in this section regarding aircraft crashes is not extensive but merely covers some basic essentials. It is intended to furnish a basic understanding of the problems involved and some basic principles generally used to cope with them. It has not been written for those departments that have specialized crash crews nor does it cover all the varied situations with which the specialized crash crews are faced. Specialized crash crews are trained not only in the basics but also in the methods used to cope with the particular aircraft that operate from their field, whether they be private aircraft, military aircraft, or the larger commercial aircraft. The knowledge and experience of the members of these crews goes far beyond what is covered in this chapter.

Aircraft crashes include those disastrous crashes where all or nearly all of the occupants are killed on impact, those forced landings where no fire exists upon the arrival of the firefighting forces, those forced landings where the aircraft is involved in fire when the firefighting forces arrive, and variations of the three.

Certain precautions should be taken when approaching crash situations regardless of the type of incident. Constant thought must be given to the possibility of a flammable liquid spill existing that could find a source of ignition during the approach. Consequently, an approach from the windward side is preferred to one from the leeward side, and one from the top of the hill is preferred to one from the bottom of the hill if a sloping terrain is involved. Drivers should also be on the watch for people who may have been thrown clear or escaped from the aircraft. This is particularly true when visibility is limited by smoke, fog, darkness, and so on, or if the response is made into an area where the ground cover is high. These factors also make it more difficult to avoid aircraft parts that may have been scattered over the area. It is good practice to lay protective lines once the apparatus has been positioned at the crash, regardless of whether or not a fire is in progress upon arrival.

Disastrous crashes. A disastrous crash includes those where all or a majority of all occupants are killed upon impact. It may be one where the aircraft dives into the ground at high speed or one where impact is made at an angle that results in parts of the aircraft being spread over a wide area. The magnitude of the problem will depend on the size of the aircraft, the number of occupants aboard, and the population density of the area where the crash takes place.

There is generally no hope for survivors if the crash is the result of a high-angle, high-speed dive. A large crater will generally be made and wreckage will be found both inside and outside of the crater. In some cases an explosion will take place and in others the crash will be followed by fire. If there is no fire, flammable liquid will be spread over a wide area, presenting the potential of a fire and/or an explosion. If there is fire in the crater, it may be advantageous to attack it with water and flood the hole. Any flammable liquid spread around the area should be blanketed with foam. It is good practice to cover the water in the crater with foam, as flammable liquid will most likely be floating on top of the water. If the first arriving units do not have foam-producing capabilities, a request for such equipment should be made immediately.

The problem is extremely compounded if the aircraft hits at a low angle and crashes into structures. Gas lines will generally be broken and electrical lines severed. Fire will normally consume part of the aircraft and a number of surrounding buildings. It is imperative that an immediate call for help be made, with the request including ambulances, police assistance, and a sufficient number of firefighting units to cope with the immediate fires and exposures and the potential for expansion. It is extremely important not to underestimate the magnitude of the situation. Most likely assistance will be coming from a considerable distance away, and it will take a long time for additional help to arrive. The delay and resultant loss will be much greater if the situation is underestimated. The incident command system for major emergencies should be placed into operation immediately.

These disasters are normally of such a magnitude and compounded by so many variables that it is almost impossible to establish any guidelines as to what action should be taken first. Of course, the saving of life and safety to personnel will always take first priority.

Forced landings—no fire. Some forced landings where no fire is involved are made with the wheels down, others are made with the the wheels up. Although the objectives and procedures for handling the two are similar, the situation is normally more hazardous when a wheels-up landing is made. The reason for this is that there is a much greater chance of fracturing flammable liquid tanks or lines when the landing is made with the wheels up.

All of the precautions listed for approaching an aircraft crash should be observed. The approach should always be made with the thought that fire could break out at any second. Consequently, lines should be laid in such positions as to make any rescues and knock down the fire if ignition takes place.

Any flammable liquid spills should be covered with foam as soon as possible, providing of course that the responding vehicles have this capability. If not, requests for units having this capability should be requested immediately.

If the first responding units do not have foam-delivery capabilities, the desired results will have to be accomplished with the the use of water. Streams will have to be used to push the spilled liquids away from the aircraft and to safe areas. Care should be taken to ensure that the spilled liquids are not pushed toward buildings or into sewers where disastrous results could occur if ignition takes place.

Every attempt should be made to stop any leaking fuel. It may be possible to stop any leak by closing fuel valves; however, if necessary the lines can be crimped or the severed ends of the lines plugged. The firefighter performing this operation should be protected at all times by a hose line.

Occupants of the aircraft should be removed when it is safe to do so. The regular exits should be used if possible. If not, second choice should be given to the emergency exits. It will probably be necessary to breach the outer skin of the aircraft to effect rescue if both the regular and emergency exits have been frozen shut by the impact of the landing. Hand or power tools can be used for this purpose.

Extreme care must be taken to prevent ignition if the spill was covered with foam and power tools or any spark-producing equipment is used for gaining entrance into the aircraft. It is possible that the protective blanket could have been broken by those carrying the cutting equipment to the aircraft. Any breaks in the protective blanket should be covered. Manned, loaded lines should be in position during cutting operations as an additional precautionary measure.

Forced landings with fire. This type of emergency is the most difficult of all. Every member of the responding companies is aware of the critical position of occupants of the aircraft and the need for quick action if lives are to be saved. The situation has most likely been caused by a fracturing of flammable liquid lines or tanks, which released flammable liquids that were ignited by sparks resulting from the impact of the aircraft with the landing surface. In many cases the fire starts before the aircraft comes to rest. The time frame is critical. Rescue efforts must be commenced within a few minutes if any success whatsoever is to be achieved.

The same precautions must be observed as for other types of crashes. The direction of the initial approach is critical. The wind direction, the terrain, and the exit facilities on the aircraft must be carefully considered. Additional help should be requested if it appears that it is or might be needed.

Although rescue is the most important factor, it must be kept in mind that a quick, direct attack on the fire may be the best means of accomplishing it. A maximum effort must be launched if the decision is made that this is the best possible method of saving lives.

It is usually best to make the initial attack with at least two lines using foam if rescue efforts are to be made while the fire is still in progress. Light water is preferred.

The most severely exposed area should be hit first and the lines directed outward to cover the remaining parts of the fuselage. (See Figure 9.3.) This action will cool down the fuselage and reduce the heat exposure to the occupants. This protective measure should be continued, with the objective of keeping the entire fuselage cool until rescue lines can be brought into play. As many lines as can be employed by the personnel and equipment on hand should be used to achieve the objective.

If the initial responding units do not have foam-delivery capabilities, the same basic procedure should be followed using water. The use of water could result in ground fires spreading. This might place firefighters in a precarious position; however, it should be kept in mind that there will probably be little chance of saving those inside the aircraft if the fuselage is not kept cool. It should also be kept in mind that no firefighter should unnecessarily be placed at risk. Protective lines should be moved into position as a precautionary measure.

If possible, the rescue team should be composed of four firefighters using spray nozzles on 1½- or 1¾-inch lines. The nozzles should be kept 12 to 24 inches off the ground and should be interlaced so as to form a solid mass of water as advancement is made on the aircraft. (See Figure 9.4.) Firefighters are provided better protection if they work inside their lines.

Forming of the team should be done as quickly as possible; however, it is most likely that the firefighters will not be positioned and lines loaded all at the same time. As each individual is positioned, his or her nozzle should be turned to a straight stream in order to help keep the fuselage cool until the entire team is formed.

It is better that the rescue team work from the windward side of the fire. This will expose the team to less radiant heat, provide better reach for their streams, and provide better visibility as the wind to their backs will blow the smoke away from them. (See Figure 9.5.)

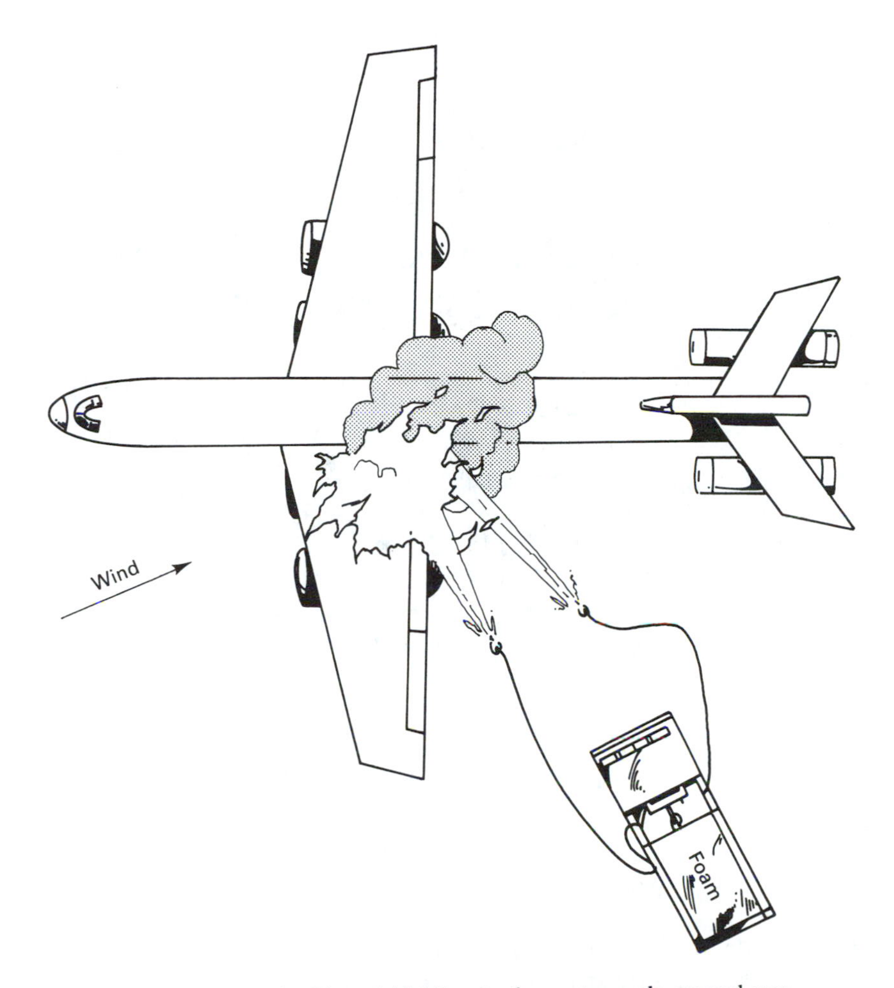

Figure 9.3 Direct initial lines to the most severely exposed area

With this arrangement of the rescue team, the two firefighters in the middle will perform the rescue work while the two outer firefighters provide them protection from the heat once the rescue exit is reached. All firefighters should be provided with full protective equipment, including breathing apparatus. If available, additional protection can be provided if the four advancing firefighters are used only for protection and two additional firefighters, who will actually perform the rescue operations, follow behind the rescue team. These firefighters should also be equipped with spray nozzles on 1½- or 1¾-inch lines and use them in the same manner as members of the advancing line. Their lines can be used to extend the width of the protective screen while helping to protect the initial four firefighters as the advancement is made. If an additional firefighter is available, he or she should follow the two to provide additional protection.

If foam is available, it is possible to put it to good use in a coordinated attack. Foam lines can be used to blanket the area behind the advancing rescue team. This

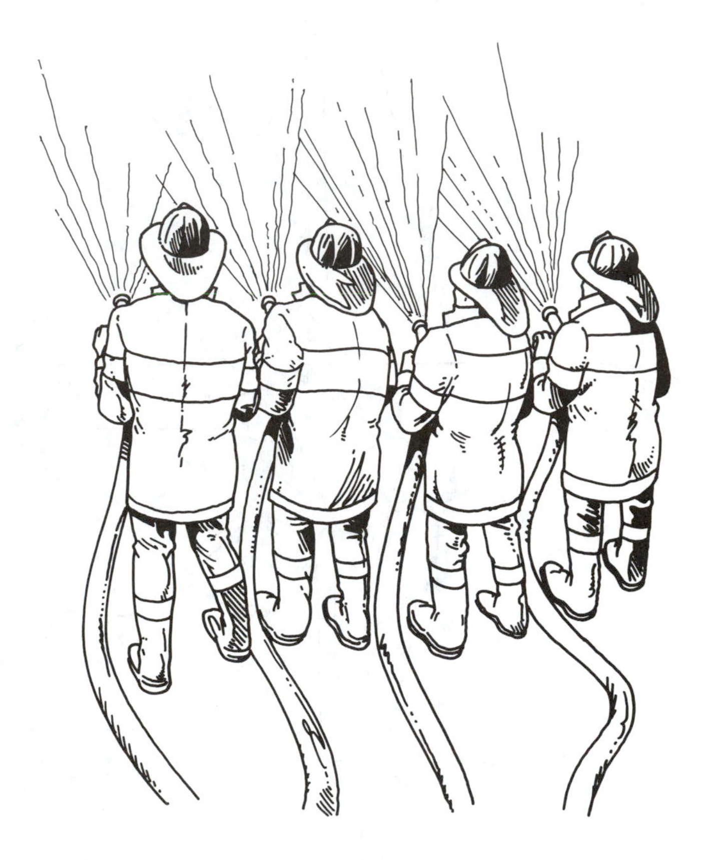

Figure 9.4 The proper positioning of rescue team members

will provide an additional protection for the firefighters pushing toward the exit doors. It is most effective when the rescue team consists of only four firefighters and the fire is being fed by a flammable liquid leak. Under such conditions there is a constant threat that the fire could get behind the advancing rescue team. Of course, backup lines are normally positioned to prevent this, but a protective blanket of foam behind the advancing team provides a greater degree of security.

It is possible for a small amount of fire to appear around the feet of some of the rescue team members if the team advances too rapidly. The natural tendency of the foam operator is to direct his or her stream at the fire when this happens. This action should be avoided, as it will most likely splash fuel onto the firefighters. It is better to direct the stream at the buttocks and legs of the firefighters and allow the foam to roll down onto the fire to extinguish it.

SMALL BOAT FIRES

The potential for a small boat fire exists in any community or fire district that has within its jurisdictional boundary a body of water of any size. Sometimes the small

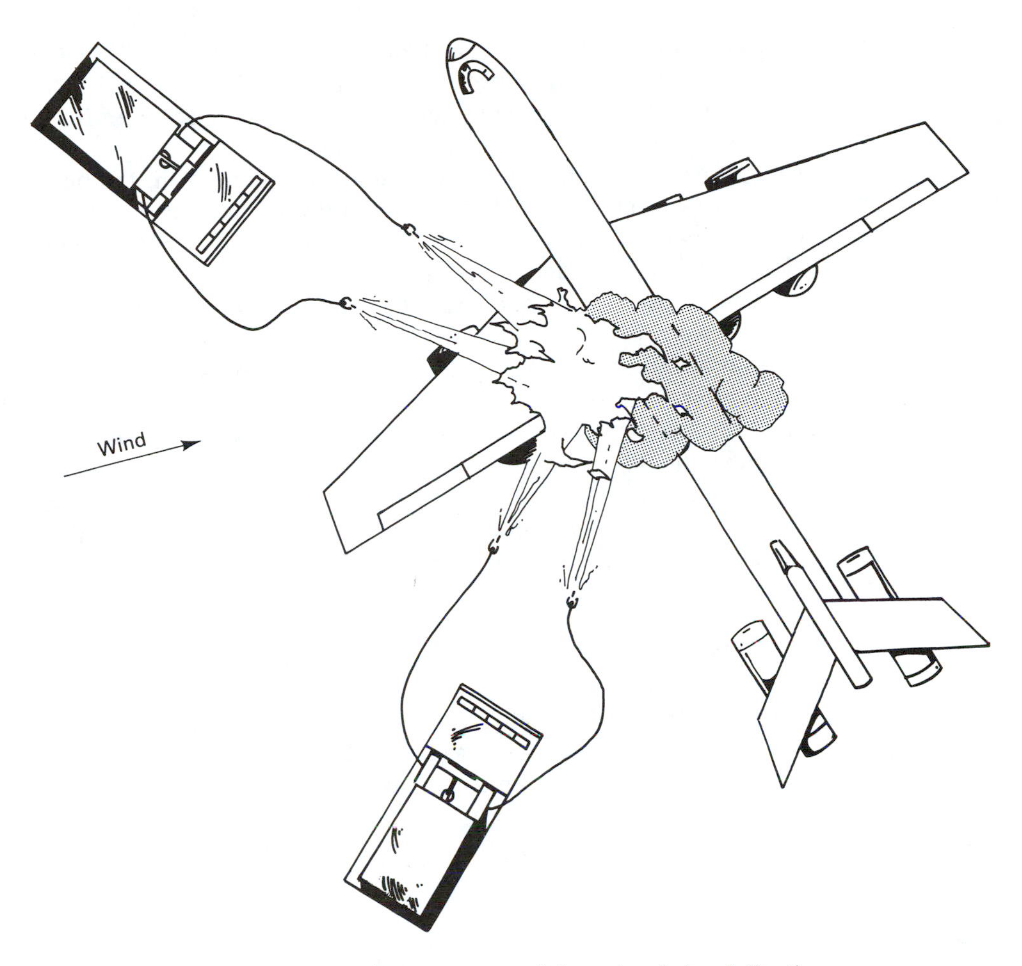

Figure 9.5 If possible, make approach from the windward direction

boats are brought to the body of water on a periodical basis, at other times they are stored permanently on the water (in most cases in marinas). The marinas will vary in size from those housing only a few boats to those housing hundreds. The storage anchorages usually consist of one or more main floating walks that extend out for varying distances into the body of water. Extending out from these main floating walks are finger piers to which the boats are secured. Many of the boats are protected from the weather during nonuse by canvas covers that extend from one end of the boat to the other. This configuration presents the constant potential for a rapid spread in the event one of the boats becomes well involved with fire.

Getting water to the fire may be a problem if the boat is tied up at a pier some distance from the shore line. Some of the larger marinas have hydrants adjacent to the main walkways and water available along the walkways. At other marinas it will be necessary to take draft to supply the hose lines. Drafting is a means of taking water from a nonpressure source such as the ocean or other open body of water.

Fires in the cabin or superstructure area should be attacked by using fog or spray streams. A single 1½- or 1¾-inch line will normally be sufficient for a fire in a small boat. Two 1½- or 1¾-inch lines will normally be required if the fire is of any size in a small yacht. A fire in a small yacht should be attacked from both sides so as not to push the fire onto an adjacent boat.

Many fires on small boats start around the engine or in the bilge area. The primary cause is the ignition of flammable liquid vapors trapped in the bilges, which is ignited when an attempt to start the engine is made. The fire usually starts with an explosion. Carbon dioxide or halon is very effective if the fire is fairly well contained under the floor plates and has not advanced too far. Light water is effective if the fire is extensive, as it has the ability to knock down the fire while forming a seal to prevent reignition. Water in the form of fog or spray streams is also effective; however, it presents the problem of possibly sinking the boat and spreading burning fuel onto the adjacent waterway. Thought should always be given to the protection of surrounding boats.

Regardless of the method used to extinguish the initial fire, there will probably be glowing embers in seat cushions, mattresses, and other combustible materials. These items will have to be thoroughly overhauled. There is also the possibility that there is a considerable amount of unburned gasoline in the bilge that will have to be siphoned into a sturdy container. It may be necessary to call for a vacuum truck if the spill is extensive. Caution should be taken to ensure that wiring is disconnected from the batteries and shore lines prior to commencing these operations. It may be necessary to pump out the boat to keep it from sinking if too much water is used.

Occasionally it will be found that a small boat on fire has been set adrift prior to the arrival of the fire department to keep it from damaging adjacent boats. In such cases the boat should be secured prior to hitting it with hose lines. If this is not done, the hose lines may push the boat into adjacent exposures.

SHIP FIRES

Fires aboard ships are considerably different from fires in structures. The problems are more unique, which presents firefighters with situations that are not common to their day-to-day firefighting efforts. The primary difference is heat conductivity. A serious fire in a windowless brick room with fire-resistive floors and ceiling would expect to be confined to the room A similar fire in a similar-sized room aboard ship would present the risk of fire spreading to adjacent compartments through bulkheads, deckheads, and decks. Bulkheads and deckheads are similar to walls and ceilings except they are made of metal, which presents the additional problem of conducted heat.

A second difference with ship fires is buoyancy. It can be expected that when master streams are used to extinguish a serious fire in a structure large volumes of water will be found running from the floors and out the doorways. The same amount of water used on a ship fire might result in the ship capsizing, as all the water would be held inside the ship. The ship's pumps might be able to dispose of part of it, but not at the rate that it is discharged by hose lines.

The third difference is the restricted capabilities for ventilation. This results in heat and products of combustion being held inside, compounding the problem of locating and extinguishing the fire.

These three differences from structure fires are common to all ship fires, but there are sufficient differences in the construction of the different types of ships so that it is impossible to establish guidelines for firefighting tactics that apply to all ships in general. Even the tactics used for fighting a fire in the engine room of a cargo ship is different than that of fighting one in the engine room of a tanker. However, knowledge of the combined tactics for fighting fires in cargo ships, tankers, and passenger ships is probably applicable to most types of commercial ships.

Cargo Ships

Cargo ships vary in size, shape, and construction features; however, there are some similarities that are common to most. Most ships are approximately 10 times longer than they are wide. The cargo is carried in holds. There are generally a minimum of five holds, numbered from the bow (front) to the stern (rear). For descriptive purposes, consider that the ship has five holds.

Number 1 hold will probably be divided into two levels, whereas the other holds will be divided into three. The three levels are referred to as the *lower deck,* the *'tween deck,* and the *upper deck*. Each of the holds will most likely be equipped with an escape hatch. An escape hatch consists of an iron ladder inside a metal tube. A small door from the tube opens into the hold at each of the various deck levels. The escape hatch may be enclosed in a small deck house on the main deck. The holds may also be equipped with ventilators, one on each side (port—left; starboard—right). The openings in the deck through which the cargo is loaded and unloaded are known as *hatchways.* The hatchways have raised sides at deck level, which are referred to as *coamings.* Coamings vary from a low level to about three feet high. The hold is secured by placing metal support beams across the hatchways and attaching them to the coamings. Hatch boards are then laid on the beams to cover the opening. Two or three heavy tarpaulins are then placed on top of the hatch boards and fastened down to the outside sides of the coamings.

To a firefighter, a hold may be visualized as a three-story warehouse that measures 60 feet by 60 feet. The warehouse is constructed of ½-inch steel plate and has no windows or doors. Similar warehouses are connected to opposite ends of the first warehouse with the ½-inch steel plate walls of the first warehousing acting as common walls between the warehouses. Each of the warehouses is filled with miscellaneous merchandise that may contain flammable or hazardous materials. None of the merchandise is skidded and there is no aisle space. The only access to each warehouse is through a 30-foot by 30-foot hole in the roof. Any additional holes that need to be cut for access or ventilation will be restricted to 15 feet from the roof on two sides of the warehouses.

The machinery space and fuel tanks will probably be found between hold number 3 and hold number 4. This is near the center of the ship. The superstructure will be located above the machinery space. The superstructure houses the crew and officer quarters, the galley, and the eating areas. The decks above the main deck in the superstructure are referred to as the *shelter deck,* the *boat deck,* and the *bridge deck,* and the *flying bridge.*

To the aft of hold number 5 is the fantail. The fantail is a deck house with extra crew quarters. Below the crew quarters will be found the steering engines, the carpenter shop, the bos'n's locker, and miscellaneous areas. (see Figure 9.6.)

The ship is compartmentalized in such a manner that one or two compartments can be damaged and become flooded without sinking the ship. For example, between each hold and between the machinery space and holds number 3 and 4 are watertight transverse bulkheads. There are no interconnecting doors or openings between the holds and none from a hold into the engine room.

Most ships are equipped with both a steam system and a CO_2 system that is capable of flooding individual holds or the engine room. Controls for these systems are normally located in the passages in the superstructure somewhere below the bridge. The controls are generally well marked.

Most ships are equipped with standpipe systems having attached 1½-inch linen hose. These systems are fed by the ship's fire pumps; consequently, they are not charged unless the fire pumps are operating.

Some areas of some ships are protected by automatic sprinkler systems. These systems have manifold inlets on the deck that can be used by the fire department to boost the pressure in the system.

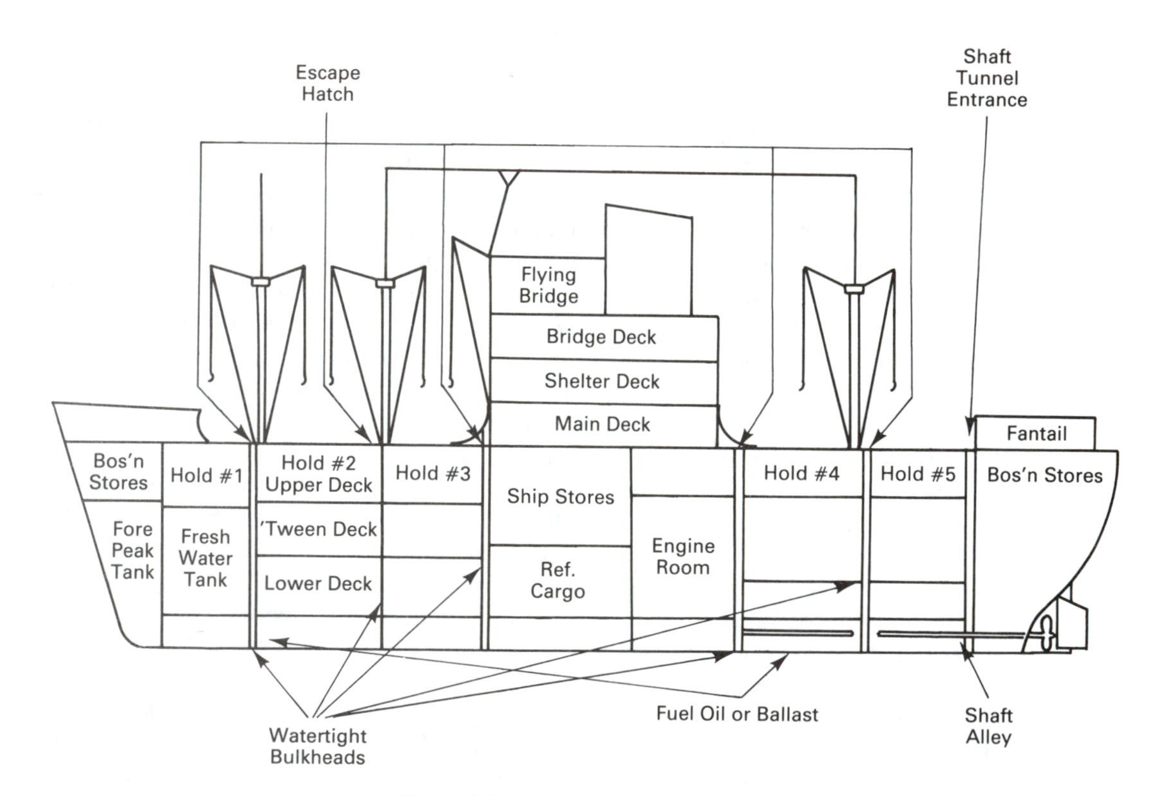

Figure 9.6 The general layout of a cargo ship

A word of caution is in order regarding the use of the CO_2 system to flood the machinery space. A warning bell normally sounds in the machinery space when this system is activated. Any firefighter in the machinery space at the time this warning bell sounds should get out as soon as possible, as the system is capable of completely flooding the machinery space in about two minutes.

Fires in the hold. It is common practice for apparatus responding to a report of a fire aboard a ship that is tied up to a wharf to drive directly onto the wharf. Smoke coming from the ship will indicate that lines will probably be needed. Consequently, the first arriving pumper should take a position close to the scene so that any equipment needed will be accessible and unnecessarily long lines will not have to be laid. Most wharfs are void of hydrants, which generally makes it necessary for the pump operator to prepare to draft. When preparing to draft, the suction hose should be dropped between the fender log and the wharf. A *fender log* is a log that is attached to the wharf at the water level to prevent approaching ships from damaging the wharf. Placing the suction hose between the fender log and the wharf will help prevent damage to the suction hose in case a boat comes alongside.

The first arriving officer should take charge of the fire and go aboard the ship to size-up the problem. It has been found best for the officer to wear a white hat rather than a helmet. The white hat appears to be a symbol of authority in marine tradition, which makes it easier for the officer to obtain needed information and the cooperation of the crew.

As with any other fire, the first job of the responding officer will be to determine the location of the fire, what is burning, and the extent of the fire. This information can generally be determined by talking with the ship's captain or first mate. The first mate will have a manifest of the ship's cargo, which will provide the information as to what is probably burning. The manifest will also indicate what materials are being carried in adjacent holds. It is important to determine if any hazardous materials are being carried and, if so, their location. The characteristics of any chemicals being carried that are not familiar to the the responding officer should be determined from the fire department's dispatch center. The first mate will also be able to provide a chart showing the layout of holds, bulkheads, ventilators, passageways, and other factors that are important to the successful extinguishment of the fire. Additionally, the first mate will be able to provide information regarding the firefighting facilities aboard ship. The ship will probably have a CO_2 system with the capability of flooding the hold. It is important to find out if this system was used prior to the department's arrival and also if the ship is being fumigated.

Once the necessary information is obtained, the officer-in-charge should notify the dispatch center of the extent of the fire and what additional equipment, if any, will be needed. The first alarm response will probably include fire boats. If the information gathered indicates that special equipment carried on the boat companies will be needed, this information should be relayed to the boat companies so that they can have the equipment laid out on deck on arrival. It will be necessary to detail several members of the land companies to help secure the fire boat lines, haul up equipment, and get a Jacob's ladder into position if it will be needed. A *Jacob's ladder* is a rope ladder with wooden steps. It is kept coiled up, which makes it easy to put over the side of the ship so that members of the boat company can climb aboard. Its construction is similar to the chain ladders developed for escape from the second floor of a dwelling.

It is best for the officer-in-charge to plan on taking full advantage of all fire-fighting systems aboard the ship such as the CO_2, steam, and foam systems. It is important, however, that department hose lines be laid out and wetted, as sometimes the ship's equipment is unreliable, particularly on some foreign ships. Hose lines should be laid from both the land companies and the boat companies.

Getting lines aboard is not always an easy task. Those from land companies are generally taken up department ladders that are raised from the wharf or hauled up by ropes that have been lowered from the ship's main deck. Lines from fire boats are generally hauled aboard by ropes lowered from the main deck. Most of the equipment from land companies is carried up the gangplank or up department ladders. Most of the equipment from the fire boats is brought aboard by lowering ropes and hauling it aboard. If ladders are used to take hose lines or equipment aboard, the ladders should be checked periodically, as the rise or fall of the tide could affect their security. A ladder should not be tied to the ship because the ship's movement could break it.

Once hose lines are in position and loaded, the hatch can be opened and the covers over the ventilators removed. Removing the hatch covers will provide additional information as to the extent of the problem. Usually the admittance of air into the hold will result in considerable smoke and heat being given off. However, if no fire is visible, it means that the first thing that will have to be done is to commence removal of the cargo.

It is important that water not be used until visible flames are encountered. Use of water at this stage will not only increase the loss but will also make the task of removing the cargo more difficult. Water will help break open any cardboard containers in the hold and certain cargo will swell, making it extremely difficult to remove.

The ship's gear, a barge crane, or a railroad crane can be used to help remove the cargo. If conditions in the hold are not too smokey or hot, it will probably be possible to load the cargo on lift boards and remove it with the use of the cranes. Longshoremen may or may not assist firefighters in this task. If conditions are such that people cannot work in the hold, then it will be necessary to remove the cargo by the use of a clam shell bucket with teeth.

Water can be used on the fire as soon as flames become visible or if the heat becomes excessive. The use of water, however, should be limited to that necessary to control the fire so that removal of the cargo can be continued. The fire may be deep in the hold or possibly in the 'tween decks. The only solution is to continue to remove cargo until the fire is reached.

Occasionally the fire may be so severe as to cause hot spots to develop on the main deck or along the sides of the ship. If left unattended, these spots can become red hot. A hot spot developing on the deck should be kept cool by a constant flow of water over it. The hot spot indicates that the main body of fire is probably directly below it. It may be possible that a hole can be cut in the deck and a spray nozzle inserted to knock down the main body of fire. The firefighter inserting the nozzle should be equipped with full protective gear, including a breathing apparatus, and should be continuously wet down with water, as the steam and heat coming from the hole will be severe. The water inserted through the hole should be restricted to the amount necessary to effect control. The nozzle should be kept in the hole in the event of a flareup, but it is best that the seat of the fire be reached by removing cargo.

Water should be played onto the side of the ship to cool down any hot spots developing there in order to prevent the plates from buckling. If a hot spot develops, it is good practice to cut a hole in the side of the ship at the hottest spot and insert a nozzle, providing of course that the hot spot is located well above the water line. A hole cut in the hull too close to the water line could cause a problem if the ship develops a list. As with cutting a hole in the deck, the personnel operating this nozzle should be continuously wetted down with a spray stream to protect them from the heat and steam coming out of the hole. It may be advantageous to increase the size of the hole to permit personnel to enter if it appears that the seat of the fire can be reached from this location. If the hole is increased to an adequate size, it may be possible to remove sufficient cargo through the hole to develop a good working area inside the hold. If this is attempted, it should be remembered that safety to personnel should always be given first consideration.

If the fire is severe and extensive, it may be necessary to use large amounts of water to gain control. This procedure, as well as that of flooding a hold, should be used only as a last resort. The overall effect is total loss to the entire inventory of the hold, but perhaps more important is the possibility that an unequal absorption of water by the cargo could cause the ship to list and possibly capsize. It should be kept in mind, however, that the responsibility for the stability of the ship is not that of the fire officer but rather that of the ship's officer who is in charge. The ship's officer should constantly be kept aware of the amount of water that is being directed into various parts of the hold. The officer-in-charge should request the ship's officer to advise him if the ship is likely to get into a dangerous condition of stability.

Thought should be given early in the fire to the possible extension of the fire to cargo in the adjacent holds by conduction through the bulkheads. If there is any indication that such a possibility exists, then the cargo in the adjacent holds should be removed. This operation should be commenced as soon as possible, as it may take 24 hours or more to empty the hold. After the cargo is removed, it is possible to cut a hole through the bulkhead to gain entrance into the fire hold; however, this is not always advisable. Many times large crates, machinery, and the like may be encountered once the hole is cut, which would limit the effectiveness of operations. It is always wise to check the manifest first to see if it can be determined as to what would be found on the fire side. Of course, if hot spots appear on the bulkhead, it would be wise to cut a hole, as this would permit direct access to the seat of the fire. All the precautions taken to provide safety to personnel while cutting a hole in the side of the ship should be taken if a hole is cut in the bulkhead.

Extinguishing a fire in the hold of a ship is not an easy task. Every piece of cargo in the hold must be removed. Cargo that has been burned should be placed in metal railroad dump cars or loaded onto trucks, wetted down, and hauled away to a place of safety. Cargo that has not been burned should be placed in a selected area of the warehouse where it can be checked for salvageable value. Although there is no indication of fire in this material, a fire watch should be established as a precautionary measure against hidden fire.

Firefighters who assist in removal of the cargo are subjected to unusual conditions. They should wear breathing apparatus during this operation even if heavy concentrations of smoke are not visible. Heated materials give off a variety of toxic fumes that are not always apparent. In addition to this, much of the work is tremendously difficult. Consequently, it is important that arrangements be made to relieve these personnel at regular intervals. It is possible that it might take several days to completely extinguish and overhaul a hold fire. In such instances, fresh crews should be brought in as often as possible.

The successful extinguishment of a fire in a hold requires careful planning and hard work. If only ordinary combustible materials are involved, the process includes five basic parts:

1. Keep the sides of the ship and the ship's deck wetted down to prevent buckling of the plates.
2. Remove the cargo until the main body of the fire can be located and extinguished.
3. Separate the burned from the unburned cargo.
4. Wet down and overhaul the burned cargo.
5. Segregate and establish a fire watch for the unburned cargo.

Hold Fires Starting at Sea. If a fire starts in the hold while the ship is at sea, the crew will generally cover the ventilators and secure the hold into as nearly an air-tight condition as possible. They will then flood the hold with CO_2 or steam and head for the nearest port. The result of their action is that the fire will continue to smolder as a sufficient amount of air is normally trapped in the hold to maintain this type of combustion.

Ships coming into a harbor with a fire reported in the hold should not be allowed to proceed to a berth until the conditions have been examined by a fire officer. If the fire officer determines that it is safe to do so, the ship should be directed to proceed to the berth that will provide the best facilities for effecting extinguishment. If the cargo is explosive or considered too dangerous, the ship should be anchored in a safe location away from congested areas. In this case, all firefighting will have to be done by boat crews and the cargo will have to be loaded onto barges for transporting to a safe location. If the cargo, together with the conditions found, indicate that the risk is too high to attempt to fight the fire, the ship should be placed in shallow water in a safe location where it can be flooded. A safe location is one where there will be no loss of life or excessive damage in the event of an explosion.

Hold Fires Containing Oxidizers. Fighting a fire in a hold containing oxidizers such as ammonium nitrate, Chili nitrate, and sodium chlorate is more a matter of knowing what *not* to do than it is to know what to do. This is one type of fire where doing the wrong thing can be disastrous. A prime example is the explosion of the S.S. *Grandcamp* in Texas City, Texas, in 1947. The cargo on this ship was ammonium nitrate. Prior to the arrival of the fire department, the hatches were battened down and covered with tarpaulins, the ventilators were covered, and the hold was flooded with steam from the ship's steam extinguishing system. From all evidence, the fire department later arrived on the scene and had prepared to put into operation a number of hose lines when the ship blew. All persons on the dock, including 27 volunteer firefighters, were killed. This explosion, together with a following explosion aboard the S.S. *High Flyer,* brought the death toll to 468. Although the ship's crew was following a traditional method of extinguishing a fire in a hold, they were not aware of the characteristics of ammonium nitrate.

Ammonium nitrate is classified as a harmless fertilizer; however, when subjected to heat, confinement, and pressure, it becomes an explosive. Unknowingly, the crew had provided all the missing ingredients. If they had followed proper procedure for fighting a fire in this type of material, most likely the explosion would never have occurred.

Ammonium nitrate, sodium nitrate, and similar materials are oxidizers. Oxidizers provide their own oxygen when heated; consequently, fires in these materials cannot be extinguished by excluding the oxygen by battening down the hatches and covering the hold ventilators. Thus, when these materials are involved in a hold fire:

1. *Do not* batten down the hatches.
2. *Do not* cover the hold ventilators.
3. *Do not* flood the hold with steam.
4. *Do not* flood the hold with CO_2

The most effective method of fighting fires where these materials are all or part of the cargo is to apply large quantities of water. The hatch and the ventilators should be kept wide open so that pressure is not allowed to build up. Serious consideration should be given to towing the ship to a safe location in shallow water if the fire is severe or if the amount of material involved is extensive. The application of large quantities of water should be continued during the towing process. The hold should be flooded once the ship is secured in shallow water.

Fire in the machinery spaces. Fires in the machinery spaces can be caused by a number of things, such as an improperly placed oil burner, a broken oil line, removal of the wrong oil strainer cover, dry firing of the fire box, and other acts of carelessness. The problem usually remains small if the fuel oil pump is shut down immediately; however, it can become serious if this action is not taken. There are a number of things that have to be done in the event of a fire. Most of them require action on the part of the crew. The induced and forced ventilation of the engine room must be stopped, the fuel pump stopped, the main engines stopped, and the skylights over the engine room opened.

Serious fires in the machinery space can present some unusual problems to firefighters. The most logical statement that can be made regarding these fires is that there is no one best method for extinguishing them. There are too many variables for any hard and fast rules to be made. There are, however, several methods that have been successfully used that can be tried.

The engine room of a ship is like a basement several stories high. The normal entrance into the engine room is through a watertight door on the main deck level. If there are indications that there is a well-involved fire in the engine room, the first lines will normally be laid to the front of this entrance. Usually the first attempt is to advance a line down into the engine room from this location; however, in most cases the extreme heat and smoke will prevent advancement. If this happens, it will then be necessary to try and make entrance into the engine room through the shaft alley.

The *shaft alley* is a fairly large tunnel that houses the propeller shaft. It extends from the engine room to the propeller. Entrance is made into the shaft alley through the shaft tunnel hatch. This hatch is found in the aft of the ship. It is a vertical, tunnel-like opening that extends from the main deck to the shaft alley.

Firefighters entering the shaft tunnel hatch should always wear breathing apparatus. This makes it a tight fit but breathing apparatus are necessary, as there is no assurance that the shaft alley is clear of smoke and toxic fumes. If the space is too small, it will probably be necessary to abandon the turnout coat while advancing the line. The turnout coat should be put on, however, before entrance is made into the engine room. A 2½-inch line and a foam line should be extended down the shaft alley hatch, into the shaft alley, and advanced to the watertight door to the engine room. It is a good idea to equip the 2½-inch line with a spray nozzle. (See Figure 9.7.)

The watertight door to the engine room can be opened from both the engine room and the shaft alley. Some of the doors swing out and some open vertically. It is easy to determine which way the door will open.

The door should be opened slowly and cautiously. Gloves should always be worn because the handle may be hot. It may be necessary to cool down the handle before it is operated. If it is possible to enter the engine room, then it is usually better to use the 2½-inch line to control the source of the flammable liquid and cool down the heated metal. The fire can then be extinguished by using either water or foam.

It is well to remember that there will probably be fuel oil tanks located behind the bulkheads that are adjacent to the engine room. The fuel oil in the tanks will probably become heated if the fire is extensive. This usually presents no hazard unless the vents become clogged. If the tanks are full and the riser in the vent pipe is short, there may be some fuel oil forced out of the tank through the vent. The amount will probably be small and the problem can easily be controlled.

If entry cannot be made into the engine room from the shaft alley, a secondary method can be used. In fact, some departments may prefer to try this method prior to attempting the method previously outlined. With this method, all doors leading into the engine room should be closed, including the one from the shaft alley. The engine room ventilators and skylights should be covered. A 1½- or 1¾-inch line should be used to keep the ventilators and skylight housing cool. Another 1½- or

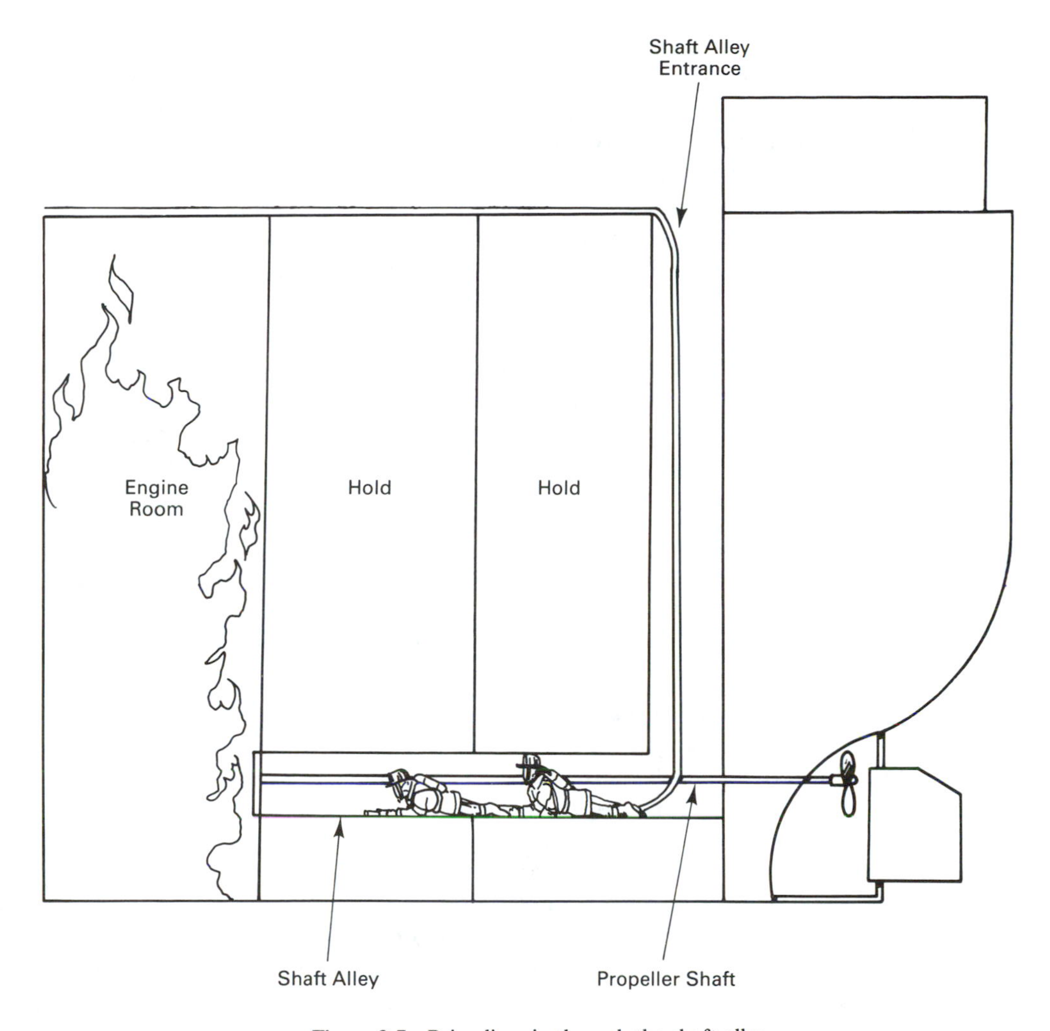

Figure 9.7 Bring lines in through the shaft alley

1¾-inch line should be advanced through the shaft alley to the opening into the engine room. Two 1½- or 1¾-inch lines equipped with spray nozzles should be lowered approximately 15 to 20 feet into the engine room through the skylight housing. It will be necessary to remove the skylight tarpaulin while this is being done, but the tarpaulin should then be replaced.

Water should be supplied to the two lines as soon as they are lowered into the engine room. It will probably take 30 to 45 minutes before the fire is extinguished and the area cooled sufficiently for the line in the shaft alley to be advanced into the engine room. The covers should be removed from the ventilators and the skylights opened prior to the shaft alley line being advanced. If this method is successful, the 1½- or 1¾-inch line will be needed only to extinguish any glowing embers in Class A material.

A third method is to attempt to extinguish the fire by using the CO_2 system.

If this system is employed, it must be done within a very short time after the fire has been discovered. Delay in operating the system may allow the metal in the immediate vicinity of the fire to reach a temperature above the ignition temperature of the burning oil. If this happens, there will undoubtedly be a reignition of the oil once the CO_2 is dissipated.

Prior to activating the system, the boiler fires should be extinguished by the crew and the auxiliaries stopped. The ventilation system should be shut down, all openings into the engine room closed, and all personnel ordered out of the engine room.

Once the engine room is flooded with CO_2, it should not be opened (except in an emergency) for some time. It is necessary to give the burning substances a chance to cool below their ignition temperature. It is good practice to take a 1½- or 1¾-inch line into the engine room and cool down the hot metal once the room is opened.

Boiler Fires. Boiler fires are not uncommon. They are usually the result of negligence on the part of those responsible for the upkeep of the boilers. Soot deposits that are allowed to accumulate in the uptakes and stacks may be loosened by mild explosions or vibration and drop onto the tops of the air heater tubes and burn. This could result in the melting of the air heater tubes. The mild explosions may be caused by the practice of lighting burners from hot brick work or inserting a torch into a gassy furnace.

The immediate action that should be taken in the event of a boiler fire is to shut down the boiler by closing off the oil supply. The burners should be removed, the air registers shut, the forced draft fan stopped, and the inlet damper closed. The water level in the upper drum should be filled. The fire can generally be located by carefully opening the access doors in the uptakes. American-made boilers usually have access doors but they are not found on all foreign ships. If available, the area should be flooded with CO_2 or halon. Water can be used if neither of these two agents is available. In fact, if the fire is advanced, water will probably be the best extinguishing agent. It should be remembered, however, that the use of water will probably ruin the inside of the firebox, but if the fire is extensive there is usually no other choice.

A couple of warnings on boiler fires are in order. One, *do not* flood the engine room with CO_2. This will not extinguish the fire and will generally hamper operations. Second, *do not* permit the soot blowers to be used to extinguish the fire, as such action might stir up a cloud of carbon dust that could ignite and explode.

Passenger Ships

Fires aboard ships that carry passengers can present a serious life problem if the passengers and crew are aboard. It is possible that the passengers could panic and all run to one side of the ship. This action is capable of capsizing the ship. A smokey fire below decks can fill the passageways with heat and smoke, making it difficult for passengers to find their way to safety and making it difficult to locate the seat of the fire.

The passenger spaces on a large passenger ship are somewhat like a hotel but present many more problems. First of all, the ship's "hotel" is made of metal. There are hundreds of small rooms and some very large rooms served by a number of passageways. The rooms on the outside are equipped with portholes, which can be used to ventilate the room but are too small to enter. The interior rooms have no direct access to the outside but are equipped with air ventilators. The head space is less than in a room in a hotel, which results in the heat condition being kept low in the event of a fire. All of these factors assist in complicating the problem.

When first arriving at a fire aboard a passenger ship, the officer-in-charge should contact the ship's officers to determine the location of the fire and the probability of what is burning. If possible, the officer-in-charge should get a member of the ship's crew to guide the firefighters. Use of 1½- or 1¾-inch lines will usually be sufficient to knock down fires in passenger staterooms, the galley, the paint locker, and so on.

Fires in the passenger spaces may be on, above, or below the main deck. If on or above the main deck, it is best that the initial lines be brought in from the main deck. Care must be used when taking these lines aboard so as not to block passenger gangplanks, which may be needed to clear passengers from the ship. It is usually best to raise aerial ladders to the bow or stern or both for the purpose of advancing these lines.

If the fire is below the main deck, it may be best to enter the ship through the cargo port on the side of the ship. These doors are at approximately wharf level and are connected to the wharf by a wide gangplank.

The age old rule of "never opening a line until the fire is seen" may have to be abandoned when advancing lines aboard ship. It may be necessary to use spray or fog streams to act as an absorbing shield to enable firefighters to reach the seat of the fire. This action will cause little damage because the bulkheads and decks in the passageways are made of metal.

Firefighters should always wear breathing apparatus and always work in pairs. The passageways are long and visibility is usually very poor, which makes it difficult for those who are not familiar with the ship to find their way around. Consequently, firefighters should not drop their line and search for the fire, as the hose line may present the only logical means for them to find their way out in the even they get lost.

All sections of the ship below, above, and to each side of the fire should be checked once the fire is knocked down. A thorough check should be made of the heating and ventilating systems. These systems extend through all decks—often through combustible partitions, false ceilings, closets and so on. The fire could extend through these systems to inaccessible spaces. A rekindle after the ship has sailed could endanger the lives of the passengers and the crew.

Tank Ships

A tank ship is constructed differently than a cargo ship. Although there is no standard tank ship, there are some features that are found on most tank ships.

On most tankers the bos'n's locker and paint locker are found in the bow.

Behind these is usually found a dry cargo hold that is used to hold miscellaneous ship's gear. Under the cargo hold are two "deep tanks" that are used to trim the ship. These tanks do not ordinarily carry oil. From the deep tanks aft are a number of storage tanks for carrying the liquid cargo. The number of these tanks will vary according to the size of the ship and its carrying capacity. The machinery spaces are located aft. The pump room will be located somewhere between the deep tanks and the machinery spaces, sometimes directly in front of the engine room. The engineer's and crew's quarters are directly above the machinery space, and the captain's and mate's quarters are located amidship under the bridge. (see Figure 9.8.)

It should be noted that the engine room is located in the stern rather than amidship. This means that there is no shaft alley that can be used to gain entrance into the engine room in the event of a fire in this location. Fighting a fire in the engine room will have to be done by using one of the other methods described for fighting a fire in the engine room of a cargo ship.

Fires in the galley, paint locker, crew's quarters, and so on aboard a tank ship are handled the same as a fire in similar spaces aboard any ship.

The most serious problem of a fire on a tanker is involvement of the flammable liquid or the potential involvement of flammable liquid. Consequently, there are certain steps that should be taken once it has been determined as to what is burning, its location, and the extent of the fire. The first is to establish a fire area. All tanks, hatches, and ports should be closed within the fire area. Ullage hole covers, Butterworth plates, tank tops, ventilator ducts, and all other vents not already closed should be closed and secured throughout the ship. Additionally, all electrical circuits in the vicinity of the fire should be de-energized. If any loading or unloading opera-

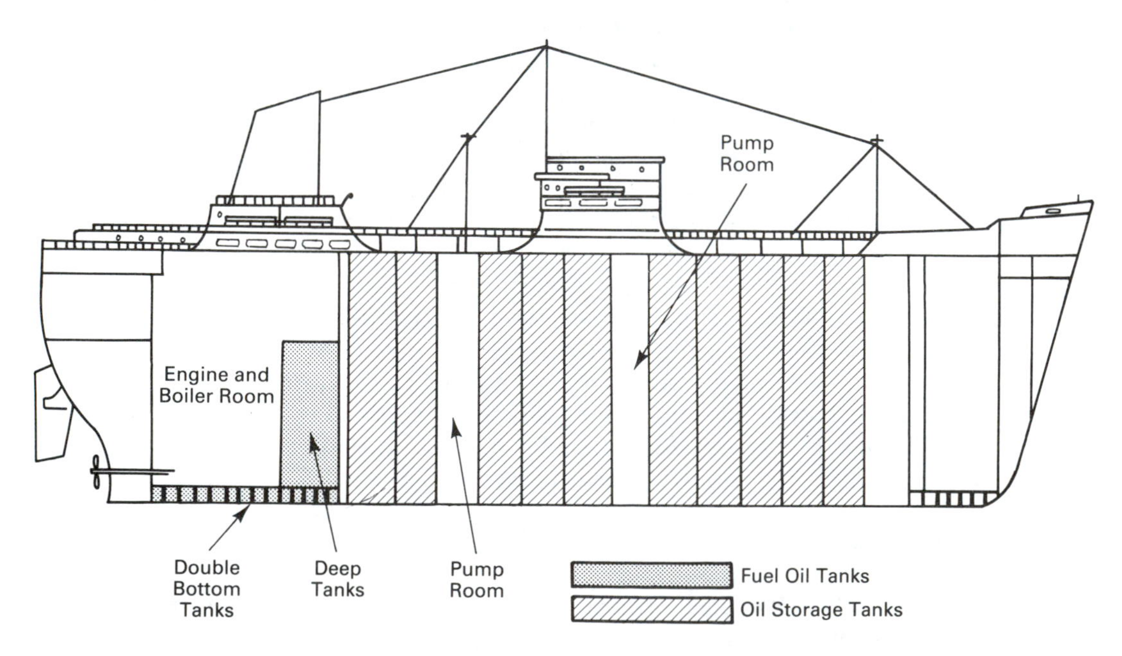

Figure 9.8 The layout of a tank ship

tions are taking place, or if any tanks are being cleaned, these operations should be closed down, all valves closed, and all hose disconnected.

Fires in cargo or bunker tanks. The first attempt to extinguish fire in these tanks should be made by excluding the air and suffocating the fire. This requires that all openings into the tank be closed. It is best that the ship's crew be used for this purpose because they are familiar with all openings to the tank. After the tank has been made air-tight, a followup that can be taken is to utilize the ship's steam-smothering or CO_2 system. The steam-smothering valves on branch lines to pump rooms, cofferdams, or tanks not on fire should be shut down prior to activating the system.

In some instances it may not be possible to close off all openings to the tank. In these cases foam will probably be the only effective agent for extinguishing the fire. Water can be used sparingly from spray or fog nozzles to pre-cool the liquid, but should not be used once the application of foam has commenced. It should be remembered, however, that water should not be applied if heavy materials such as asphalt or tar are involved, as this could result in a very severe and drastic explosion.

Pump room fires. Fires in the pump room usually occur in the bilge or somewhere near one of the pumps. The fire may be fed by a break in a line or the rupture of the seal on a pump. Sending personnel down into the pump room to extinguish a fire is an extremely dangerous practice. While they may be able to extinguish the fire, there is always the danger of a reflash. It is much safer to fight the fire from the main deck.

The first attempt should be to extinguish the fire by lowering one or more hose lines down into the pump room through the pump room hatchway. These lines should be equipped with spray or fog nozzles. If the fire is of such intensity that a close approach cannot be made to the entrance to the pump room, then the ship's built-in extinguishing system should be used. (See Figure 9.9.)

If one of the ship's systems is to be used, then it is first necessary to close all openings into the pump room and shutdown the mechanical ventilating system. If the ship is fitted only with a steam-smothering system, this system should be activated. Valves to branch lines leading to cofferdams or tanks not adjacent to the pump room should be closed. If the ship is fitted with a CO_2 water fog or foam system, one of these systems should be activated in lieu of the steam-smothering system. The steam-smothering system should be turned on in adjacent tanks and cofferdams.

As a last resort, the pump room can be flooded. This will result in either extinguishing the fire or bringing it up to the level where it can be fought from the main deck level.

Fires on the deck. Fires on the deck can be the result of a tank overflowing or they can be fed by a broken transfer hose or leaking pipeline. If the fire is being fed from a supply source, the flow of fuel should be stopped if at all possible. Regardless of the source of the fire, all tank openings throughout the vessel and the pump rooms should be closed as quickly as possible.

Figure 9.9 One attack method for a fire in the pump room

If possible, fires on deck should be attacked from the windward side. This direction of attack will have a twofold benefit. First, it will carry the flames, smoke, and fumes away from the attacking firefighters. Second, it will help carry the extinguishing agent into the fire.

Normally the best extinguishing agent will be foam. AFFF (light water) is preferred. The foam should be played against the nearby vertical structures and be allowed to run down and flow smoothly over the fire. It can also be applied to the deck ahead of the fire.

Foam is not particularly effective on flowing liquid fires, as the movement of the liquid will prevent the formation of an air-tight foam blanket. Under these circumstances, if the supply of fuel cannot be shut off and it is burning at its source, portable CO_2, dry chemical, or water fog will probably be effective when applied directly on the fuel as it emerges from the opening. It is good practice to play foam ahead of the flowing fuel to form a dam and thereby prevent the spreading of the fire. Water from spray or fog nozzles can be used to cool the surrounding structures of the vessel and to protect the firefighters using the foam. The water should, however, be used in such a manner as not to affect or break up the foam blanket or dam.

REVIEW QUESTIONS

CAR FIRES

1. What is the primary problem facing firefighters in a car fire?
2. What is one of the first precautionary measures that should be taken at a car fire?
3. How should a fire under the hood of a car be extinguished?
4. Under what conditions is it good practice to disconnect the battery on a car fire?
5. How can fires inside the car normally be extinguished?
6. What precautionary measure should be taken before opening a trunk of a car to check for fire extension?
7. How should the attack be made on a fully involved car?
8. What precautionary measure should be taken if vehicle extrication procedures have to be used to remove the occupants of a car?

TRUCK FIRES

1. What is the primary problem involving truck fires?
2. What should be done prior to overhauling a truck fire?

AIRCRAFT FIRES

1. What should be the first attempt to extinguish a fire in a piston engine of an aircraft that is confined within the engine nacelle?
2. What action should be taken if this method fails?
3. How are fires within the combustion chamber of jet engines best controlled?
4. How far should personnel remain from the front, side or directly to the rear of a jet engine intake ducts?
5. What is the minimum distance that personnel should remain from the exhaust outlet of a running jet engine?
6. What method should be used to extinguish a fire outside of the combustion chamber of a jet engine but within the engine nacelle if the fire cannot be extinguished with the aircraft's built-in extinguishing system?
7. What area should be avoided by the apparatus driver when responding to a wheel fire in an aircraft?
8. How should overheated brakes on an aircraft be handled?
9. What extinguishing agent is recommended for a wheel fire if actual flames are observed?

10. On large commercial aircraft, where is the most stable point to initiate an attack on a fire inside the aircraft?
11. What method should be used to extinguish a fire inside an aircraft if it appears that conditions are ripe for a backdraft?
12. What is the primary objective of fighting a fire in an aircraft?
13. What are some of the precautions that should be taken when approaching an aircraft crash?
14. From which direction is it generally best to make the approach to an aircraft crash?
15. How should the fire in a hole caused by a high-angle, high speed crash of an aircraft be extinguished?
16. What action should be taken on a forced landing of an aircraft where no fire is involved?
17. How can the leaking fuel on an aircraft be stopped?
18. What precautionary measures should be taken before cutting into an aircraft for the purpose of making rescues?
19. What is usually the most effective method of saving lives at an aircraft fire?
20. If rescue efforts are to be made while an aircraft is on fire, what is usually the best method of making the initial attack?
21. What extinguishing agent is preferred for aircraft fires?
22. When making an initial attack on an aircraft for the purpose of making a rescue, which area of the aircraft should be hit first with the foam lines?
23. If possible, what should be the composition of a rescue team that is to be used to force an opening through the fire to the aircraft door?
24. Which firefighters in the rescue team perform the rescue work?
25. How can foam lines be used to protect the rescue team?

SMALL BOAT FIRES

1. How should a fire in the cabin or superstructure area of a small yacht be attacked?
2. What is the primary cause of fires on small boats that start around the engine or in the bilge area?
3. What extinguishing agents are effective if the fire around the engine of a small boat is contained under the floor plates and has not extended too far?
4. What action should be taken in a small boat fire if there is a considerable amount of unburned gasoline in the bilge after the fire is extinguished?

SHIP FIRES

1. What is the primary difference between a fire aboard ship and one in a structure on land?
2. What is the second difference?
3. What is the third difference?
4. Approximately how much longer is a ship than it is wide?
5. What are the three levels of decks in the hold of a ship?
6. What is the construction of the escape hatch on a hold?
7. What is generally found in the superstructure of a cargo ship?

8. What is the fantail on a cargo ship?
9. Where are the controls for the steam and the CO_2 systems generally found on a cargo ship?
10. Where should the first responding pumper to a cargo ship fire with smoke showing generally be spotted?
11. Why should the first arriving officer at a ship fire wear a hat aboard ship rather than a helmet?
12. What is the first job of the first arriving officer?
13. Who will probably have the manifest of the ship's cargo?
14. What will the manifest tell the first arriving officer?
15. What precautionary measure should be taken if it is decided to use the ship's firefighting systems to attempt to extinguish a fire in the hold?
16. How are hose lines taken aboard?
17. When should the hatch be opened?
18. When should water be used on a fire in the hold?
19. How is the cargo removed from the hold?
20. What action should be taken when hot spots develop on the deck?
21. What action should be taken when hot spots develop on the side of the ship?
22. When should the hold of a ship be flooded if it contains regular cargo?
23. Who is responsible for the stability of the ship during firefighting operations?
24. What precautionary measure should be taken to ensure that fire has not spread by conduction to the cargo in adjacent holds?
25. What should be done with the cargo that is removed from the hold?
26. What are the five basic parts to the successful extinguishment of a fire in a hold that contains ordinary combustible material?
27. What action is normally taken by the crew if a fire occurs in the hold while the ship is at sea?
28. What action should be taken by the officer-in-charge regarding a fire coming into the harbor when the fire started in the hold while at sea?
29. What is the classification of ammonium nitrate?
30. What three conditions will change ammonium nitrate to an explosive?
31. What four things should *not* be done when fighting a fire in a hold that contains oxidizers?
32. What is the most effective method of fighting a fire in the hold when the cargo is ammonium nitrate?
33. There are a number of things that should be done in the event of a fire in the machinery spaces of a ship. Most of these require action on the part of the crew. What are these things?
34. To what point should the first lines normally be laid for a fire in the engine room?
35. Where should entrance to the engine room be made if it cannot be made through the door on the main deck?
36. What is the shaft alley?
37. Where is entrance made into the shaft alley?

38. What lines should initially be extended down the shaft alley and to the engine room entrance?
39. How is the engine room door from the shaft alley opened?
40. What hazard do the fuel oil tanks located behind the bulkheads in the engine room present?
41. What method of extinguishment can be tried if entrance cannot be made into the engine room from the shaft alley?
42. What size lines should be lowered into the engine room using this method?
43. When should the CO_2 system be used to try to extinguish a fire in the engine room?
44. What should be done prior to activating the CO_2 system?
45. What is the immediate action that should be taken in the event of a boiler fire?
46. How can the fire in a boiler generally be located?
47. What material should be used to extinguish a boiler fire?
48. What are the two "don'ts" for a boiler fire?
49. What should the officer-in-charge do when first arriving at a fire aboard a passenger ship?
50. What size lines are usually necessary to knock down fires in staterooms?
51. What is the best method of bringing lines aboard ship if the fire is above the main deck on a passenger ship?
52. What is the best method of bringing lines aboard ship if the fire is below the main deck on a passenger ship?
53. What age-old firefighting rule may have to be abandoned when advancing lines aboard ship?
54. Where should checks for extension of the fire be made if the fire was in a passenger stateroom?
55. Where are the bos'n's locker and paint locker found on a tanker?
56. Where are the crew's quarters located on a tanker?
57. Where is the engine room located on a tanker?
58. What is the most serious problem of a fire on a tanker?
59. What things should be done once it is determined what is burning on a tanker?
60. What should be the first attempt to extinguish a fire in a tank on a tanker?
61. What should be done after the tank has been made air-tight?
62. What should be the first attempt to extinguish a fire in the pump room aboard a tanker?
63. What should be done if one of the ship's extinguishing systems is used to extinguish a fire in the pump room?
64. What should be done as a last resort for a fire in the pump room?
65. What is generally the best extinguishing agent to use to extinguish a flammable liquid fire on the deck on a tanker?
66. What method should be used to extinguish a flowing liquid fire on the deck of a tanker?

CHAPTER 10

Miscellaneous Fires

OBJECTIVE

The objective of this chapter is to introduce the reader to firefighting tactics as they apply to flammable liquid fires, flammable gas emergencies, electrical fires, brush fires and pier and wharf fires. Review questions are provided at the end of the chapter to assist the reader in determining his or her level of comprehension of the chapter contents.

The title to this chapter is somewhat misleading. It almost sounds as if the firefighting practices to be discussed are of less importance than those outlined in previous chapters. This is not the case. Many of the fires covered in this chapter are extremely challenging to firefighters. In fact, a number of firefighters have been killed and others injured during the violation of some of the basic safety principles outlined here. The information contained in this chapter should be thoroughly understood by those involved in the practice of firefighting—not only for the purpose of doing a more effective job in fire extinguishment but also for the purpose of establishing practices of personal safety.

FLAMMABLE LIQUID FIRES

From a technical standpoint, a flammable liquid is a liquid that has a flash point below 100° F. Those liquids having a flash point of 100 degrees F. or more are referred to as *combustible liquids*. However, for the purpose of discussing firefighting principles, both will be considered as flammable liquids. Oil, however, will be excluded from this definition.

As discussed in Chapter 1, flash point refers to the point at which a flammable liquid will start giving off sufficient vapors to burn. Consequently, it can be said that vapors will be given off that will burn any time the flash point is below the ambient temperature. Those liquids having a flash point above the ambient temper-

Photo 10.1 Some of the barrels stored in this building projected as far as 230 feet. (Courtesy of Wichita, Kansas, Fire Department)

ature will have to be heated before they will give off vapors that will burn. In some cases, this concept is a factor that should be considered when determining the way in which a fire in a flammable liquid will be attacked.

Another concept of flammable liquids that has an impact on firefighting procedures is whether or not the flammable liquid is *miscible.* If it will mix with water, it is miscible; if it will not, it is not miscible. A good portion of the flammable liquids, including gasoline, encountered in firefighting are not miscible. These liquids are referred to as *hydrocarbons.* The overall effect is that if water is used as the extinguishing agent on a hydrocarbon fire, the flammable liquid will float on top of the water and continue to give off vapors once the fire is extinguished.

Those flammable liquids that are miscible are referred to as *polar solvents.* Water mixed with polar solvents will increase their flash points and reduce the hazard. However, conventional foams cannot be used to extinguish fires in these liquids because the liquid will destroy the foam. Alcohol is an example of a polar solvent. Polar solvents require special foams for extinguishment.

Some flammable liquids are a mixture of hydrocarbons and polar solvents. Two examples are unleaded gasoline and gasohol. These mixtures present special problems.

Spill Fires

One characteristic of a flammable liquid spill that finds a source of ignition is that the resultant fire will propagate over the entire spill area at an amazing rate of speed. Consequently, there are special precautions that should be taken when approaching a flammable liquid spill fire, or for that matter, a flammable liquid spill that has not yet become ignited. If possible, approach the fire or spill from the

Photo 10.2 Foam being used on a flammable liquid spill from an overturned gasoline truck. (Courtesy of Chris Mickal, New Orleans Fire Department Photo Unit)

windward side and from the uphill side if the terrain is sloping. This is particularly important if ignition has not yet taken place. The vapors capable of being ignited may be invisible, which could place the apparatus in the vapor area or possibly in the spill area itself if approach is made from any other direction. It does not take much imagination to visualize the potential result if ignition occurs while the apparatus is so located.

When referring to a flammable liquid fire, the size of the fire is determined by the surface area involved, not the amount of fuel available to burn. If the surface area is small, the fire is small. If the surface area is large, the fire is large.

If the fire is small and not being fed from a continuous source, it may be possible to extinguish it with a few shovels of loose dirt. If dirt is not available, or the fire is too large to handle in this manner, a carbon dioxide or dry chemical extinguisher may prove to be effective. As a general rule, a carbon dioxide extinguisher can be used on fires up to about 25 square feet of area, whereas a dry chemical extinguisher may be effective up to about 100 square feet of area. The dry chemical extinguisher is preferred in either case because it is not only capable of knocking down the fire quicker but it also has the advantage of being effective at a greater distance from the fire.

Fires larger than 100 square feet should, if possible, be attacked with foam. The most useful foam for hydrocarbon flammable liquid fires is light water (AFFF). Polar solvents require special foams.

The attack should be made by directing the foam ahead of the fire and allowing it to spread over the fire. If vertical surfaces are available behind or to the side of the fire, the foam can be projected onto these surfaces and allowed to run down and out onto the fire.

An even more effective method of knocking down a large flammable liquid fire is the combined used of dry chemicals and foam. Of course this method is limited to those departments that have the capability of discharging large quantities of dry chemicals on a fire. The attack on the fire should initially be made with dry chemicals, which could possibly result in complete extinguishment of the fire. This attack should be followed immediately by completely covering the flammable liquid with foam to prevent a reignition. Failure to follow through with the foam blanket could result in a more critical situation than the original fire.

If foam is not available, it may be necessary to extinguish the fire by using water. Water can have a dual effect, depending on the flash point of the material involved. The water may have the effect of cooling the liquid below its ignition temperature or it may be used to separate the liquid from the flame. A general rule is that those liquids having a flash point above 100° F. can possibly be extinguished by cooling, whereas those having flash points below 100° F. cannot.

Attacking a spill fire with water will result in the fire being pushed away from the direction of attack. For example, if the fire is attacked from the north, the fire will be pushed toward the south. It is important that the safest direction for pushing the fire be determined prior to commencing the attack. If exposures are involved, care must be taken to see that lines are laid to protect them prior to the attack being initiated. It should also be remembered that almost all hydrocarbon liquids are lighter than water. Consequently, any unburned liquid will float on top of the water after the fire has been extinguished. This could present the potential hazard of a reignition.

One of the better methods of making the attack on the fire is through the use of at least two spray nozzles on 1½-, 1¾-, or 2½-inch lines backed up by one or more similar protective streams. (See Figure 10.1.) The size and number of lines required will depend on the size and extent of the fire. The attack should be made from a direction that will push the fire to the least hazardous area.

A careful analysis should be made prior to the attack as to the amount of water available. The attack should not be started if there is not a sufficient amount of water available to complete the job. If extinguishment cannot be completed, the overall effect would most likely be a spread of the fire, which would intensify the overall problem.

In some cases, the most effective method of handling a spill fire is to let it burn itself out. This method would most likely be used in those instances where there is doubt that the water available can do the job and where no exposures are involved.

In many situations, the spill fire is being fed from a continuous supply source. The magnitude of the problem will be affected by what is burning, the size of the fire, the potential for its spread, and the number and type of exposures. The solution to the problem will include the need to protect the exposures, the supply, stop the spread, and extinguish the fire.

The exposures could include a variety of types, including storage tanks or other flammable liquid facilities. (Exposure-protective tactics for storage tanks and

Figure 10.1 A good method of attacking spill fires

other flammable liquid facilities will be discussed later in the chapter.) Standard procedures for protecting exposures should be employed for other types of exposures. The key is to use a sufficient number of lines to complete the task effectively. In some cases it may be necessary to use hose lines to divert the fire to a safer direction of extension.

Shutting off the fuel supply to the fire should be given early consideration. Shutting off the flow generally requires that a valve be closed or a hole plugged. This action requires a coordinated effort by several firefighters. The team assigned the task of shutting off the flow should consist of a minimum of three firefighters. The two outside firefighters should man 1½- or 1¾-inch lines equipped with spray tips. The inside firefighter is usually an officer. This person is responsible for making the shutdown. When advancing, the inside person maintains the integrity of the team by keeping one hand on the inside shoulder of each of the nozzlemen. (See Figure 10.2.) The advance should be made from the direction that will best accom-

Figure 10.2 A typical approach to shut off a valve

plish the task while providing the maximum safety to the firefighters. The wind direction, the terrain, and other factors that will influence the potential spread of the fire and the safety to personnel should be kept in mind when making the decision as to the direction of attack. It is good practice to provide a backup line as a safety factor for the advancing shutoff team.

Stopping the spread of the fire caused by flowing fuel can become a tricky job. The ideal is to dam up the area directly in the path of the flow direction. The heat given off by the fire is normally too intense to permit working close to the fire's edge. In these instances it is best to work some distance ahead of the fire and to use sand, dirt, or other noncombustible material to provide a dam. It is a good idea to request several loads of dirt or sand through the dispatch center if a sufficient amount is not at the scene of the emergency. It may be possible that the flow can be shut off before the requested material arrives on the scene; however, it is also possible that the source of the flow cannot be stopped. It is much better not to need the material than to suffer the consequences of the fire spread. Incidentally, if the fire is located close to a road, it is not a bad idea to place a firefighter there to watch for truck loads of sand or dirt that might be in transit to a construction site. Confiscating this material could prove to be extremely beneficial to the control of the flow problem.

If it becomes impossible to stop the flow while attempting to shut off the supply, it may be necessary to divert it to a safer area. Shovels can be used ahead of the fire to provide flow paths, or hose lines can be used to change the direction of the flow.

Extinguishing the fire can be done simultaneously with protecting the exposures, stopping the flow, and stopping the spread, or at times it may be necessary that extinguishment be delayed until some of the other requirements are fulfilled. Regardless, extinguishment operations should be conducted as previously discussed.

Tank Fires

The problems of fighting fires in flammable liquid storage tanks will vary depending on the standard variables of weather, tank size, material burning, and so on. Each of these must be considered when making a size-up of the fire. However, when considering the problem of exposures, there is a considerable difference in the hazard when the exposures are horizontal storage tanks than there is when the exposures are vertical storage tanks.

Generally, a fire threatening a flammable liquid storage tank is the result of a spill from a storage tank or transportation line becoming ignited, or from a fire in an adjacent tank. Consequently, the problem is initially the same regardless of whether the exposed tanks are vertical tanks or horizontal tanks.

The problem will primarily be one of protecting the exposures and extinguishing the spill fire if the fire is confined to a diked area. Normally, the exposure problem will be eliminated once the spill fire is extinguished. However, the problem of protecting the exposures may be continuous if the fire is being fed by a break in a supply source or as a result of a fire in an adjacent tank. In the case of a break in a tank, the tank feeding the fire in the diked area is itself an exposure.

If the liquid has pushed its way out of the diked area, or no diked area is involved, the problem becomes one of protecting the exposures and extinguishing the spill fire. In some cases, the spill area will include pipe trenches.

The extinguishment of spill fires that expose storage tanks follows the same basic principles for extinguishing spill fires where storage tanks are not involved. The basic difference is the extinguishment of fires in the tanks themselves and the protection of exposures.

Many times the spill fire is so large that it is impossible to extinguish it with the foam carried on a single engine company. In this case it may be better to cool the tank or exposure until a sufficient amount of foam is on the fireground to extinguish the fire. If it is necessary to depend on the foam carried on engine companies, then the situation may require several engine companies to accumulate a sufficient amount of foam to do the job.

Vertical storage tanks. Fires may occur in vertical tanks or the tanks may be exposed to spill fires. The best method of extinguishing the fire in the tank will depend on what is burning, the size of the tank, and the construction of the tank top. If the roof of the tank is gone, it may be possible to extinguish the fire by using water if the tank is small and a hydrocarbon liquid is involved. This is particularly true if the entire surface area of the fire can be covered with water fog at the same time. Care has to be taken, however, to limit the amount of water, as too much water could cause the tank to overflow. If a considerable period of time has elapsed between the ignition of the fire and the application of water, it will be necessary to cool down the inside exposed surfaces of the tank to prevent reignition.

If the tank is too large to extinguish the fire by using water, the application of foam will prove most effective. If the tank has its own foam-extinguishing system, it should be used. If the system is not available on the tank, then foam applicators will have to be employed. Although it is possible to apply the foam through hand-held hose lines, in most cases this is neither practical nor feasible.

In many cases it is not possible to extinguish the fire and it has to be left to burn itself out. Occasionally it is possible to start discharging the fuel in the tank to other storage tanks through the tank's distribution system. This will cut down the burning time and reduce the overall fire loss. The firefighting problem becomes one of protecting the exposures and keeping the burning tank cool whenever a fire is allowed to burn itself out.

It is important to understand the construction of a vertical tank to protect it adequately as an exposure. Vertical tanks are usually designed so that the weakest part of the tank is the roof. The roof of the tank will normally give way under a pressure buildup within the tank. This action will send the roof skyward. The distance it will travel will depend on the degree of the pressure buildup and the size of the roof. This action could rupture adjacent tanks and expand the overall problem. It always presents a serious hazard to firefighters.

Records show that on a number of occasions the base of the tank rather than the roof releases on a pressure buildup. This action sends the entire storage tank into the air, causing it to react as a missile.

If the tank has a dome roof and is burning at the vent, it is usually a simple task of extinguishing the vent fire. This may be accomplished by using dry chemicals or water in spray or fog form.

Fortunately, most of the modern vertical storage tanks have floating roofs. Fires in these tanks generally occur in the seal area between the tank and the floating roof. These fires can generally be extinguished through the proper use of light water or foam. Firefighters must be extremely cautious if an attempt is made to extinguish these fires with the use of water. It is possible to sink the floating roof. This will usually result in a simple fire developing into a fully involved tank fire with its magnitude of problems.

The hazard to an exposed tank depends to a considerable degree on the amount of fuel in the tank. The hazard is much greater to a partially filled tank than it is to a full one. The principle involved might be illustrated by the use of a paper cup holding water. The water in the cup can be heated to its boiling point from an outside source without damaging the cup as long as the heat is applied to an area of the cup containing liquid. The water absorbs the heat and the cup remains intact. However, once the water boils away, the fire will heat the vapor in the cup, the cup will absorb the heat and the cup will ignite.

Fires in flammable liquid storage tanks may burn for several days if allowed to burn themselves out. Large amounts of water should be applied to the burning tank and to all exposed tanks during the entire burning period. It may be necessary to utilize all hand-held lines; however, if there is an adequate water supply it is wise to set up portable monitors both for protective streams and to cool the tank on fire. This not only relieves the strain placed on firefighters but makes it possible to continue to protect the exposures in the event that a manned position has to be temporarily abandoned due to an unexpected flare-up or other emergency.

Horizontal storage tanks. For the purpose of this discussion, the term *horizontal storage tank* includes tank trucks and railroad tank cars as well as permanently installed tanks.

Fires seldom occur in horizontal storage tanks. Occasionally, fires will occur at the vents, but extinguishment of these fires is usually a simple matter of cooling down the tank. This action will reduce the pressure buildup in the tank and close the pressure relief valve. The most pressing problem with horizontal storage tanks occurs when they are directly exposed to a fire. The exposure problem to these tanks is much more critical than it is to vertical tanks.

The weakest part of these tanks is the ends. There is a possibility that the ends will give way if pressure builds up within the tank faster than it can be released. The tank literally becomes a rocket when this happens if the released fuel is ignited as it escapes. The tank will be projected a considerable distance, with fire flaming out the rear end. Two examples are worth noting.

A number of years ago in Kansas City, Kansas, the end of a horizontal tank gave way while firefighters were fighting an extensive flammable liquid fire. The reaction caused the tank to break from its mounting and to rocket 94 feet ahead. During its progress it went through a 13-inch brick wall and also knocked down another wall of a brick service station building. The fire shooting out the rear of the tank cost the lives of five firefighters and one civilian.

Photo 10.3 Tank car fires should be considered as horizontal storage tanks. (Courtesy of Chris Mickal, New Orleans Fire Department Photo Unit)

Another incident occurred in Pennsylvania—this one involving a tank truck. When the ends of the tank let go, the remaining parts of the tank and truck were projected approximately 470 feet down the road where the tank smashed into a stone retaining wall. Parts of the truck cab and running gear were torn loose during the progress. After hitting the wall, the tank tumbled and rolled over the highway and through a crowded intersection. It came to rest approximately 900 feet from its original position, leaving eleven people dead in its path of travel.

A horizontal tank can become exposed to fire as a result of a fuel spill or from a fire in an adjacent tank or structure. If it is due to a fuel spill, the spill fire can be extinguished by covering it with foam. However, if the spill fire is being fed by a break in the tank or from another source, top priority will have to be given to stopping the leak and pushing the fire away from the exposed horizontal tank. Tactics explained in the section on spill fires can be used for this purpose. Application of foam will have to wait until the leak is eliminated if the source of the leak is the horizontal tank. The reason for this is that the water used for the advancement of the shutoff team will wash away the protective foam blanket. It may be possible to use water to wash the spill to a safe area.

It is important that water be used to cool a horizontal tank any time it is being exposed to fire. It will normally take a minimum of 500 GPM applied to an exposed side to prevent a pressure buildup. Either hand-held lines or heavy streams can be used for this purpose.

A basic principle that should not be ignored when playing water on an exposed horizontal tank is *never direct the water from the ends of the tank.* To do so would place firefighters in extreme jeopardy if an end of the tank let go. (See Figure 10.3.)

It is possible that a tank truck or railroad tank car can be involved with fire in an area where there is an insufficient supply of water to keep the tank cool, and a sufficient amount of foam is not available to extinguish the fire. In these cases it

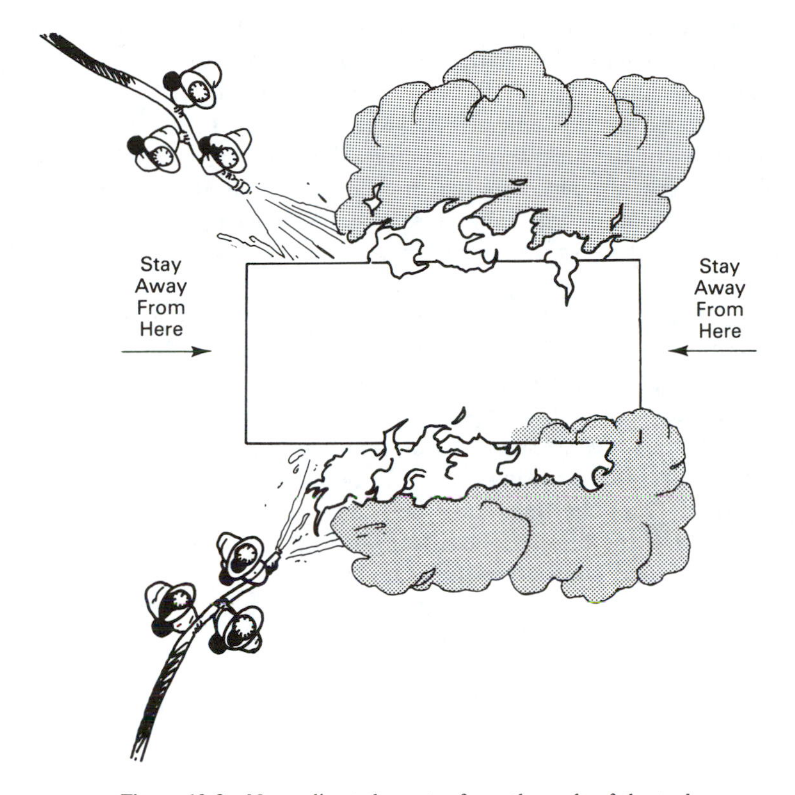

Figure 10.3 Never direct the water from the ends of the tank

will be necessary to evacuate the area. This means that everyone within a distance of at least 3,000 feet in all directions should be removed. All available firefighters and police officers should be used to accomplish this task. If any amount of water whatsoever is available and if time permits, it may be wise to set up an unmanned portable monitor to help keep the tank cool while the evacuation is taking place. This may eliminate the potential explosion, which would allow the fire to harmlessly burn itself out.

Fires Resulting from Breaks in Transportation Lines

Occasionally flammable liquid spill fires are fed by breaks in transportation lines. At times these lines are above ground, but in most cases they are underground lines that have been broken by construction equipment. The extent and seriousness of the fire will depend primarily on the area covered by the spilled liquid prior to ignition and the rate at which the spill is being fed. The basic procedure for attacking spill fires should be used to extinguish the fire. The problem then becomes one of stopping the flow. The problem is somewhat simplified if a valve is readily available that can be closed to stop the flow. The method of making an approach to the valve was previously explained.

Unfortunately, in most cases when the break is in an underground line the nearest valve for shutting down the line will be located at a remote location and will have to be shut down by the company owning the line. Determining who owns the line is at times difficult, as several pipes used for transporting flammable liquids may run down the same street. In these circumstance it will be necessary to relay information to the dispatch center as to what type of liquid is involved and the distance that the pipe line is from the curb or another identifiable reference point. This information will help the dispatcher determine the owner of the line. In the meantime, there is some action that can be taken that will help minimize the problem.

If the transportation line is small and the flammable liquid flow from the line is minimal, a single 2½-inch line can be advanced to the break. For larger transportation lines and greater flows, two or more 2½-inch lines may be required. Firefighters advancing the lines should be protected at all times by large quantities of water from spray streams. Additional lines should be used to channel the escaping liquid to a safe location. It may be necessary to build dams to confine or control the flow. Once the lines have been advanced to the break, the nozzles should be opened and the lines pushed deep into the pipe. Where possible, positive shutoff nozzles should be used to eliminate the possibility of the nozzle accidentally being shut off after the line is inserted into the pipe. The objective is for the water to push the flammable liquid back into the pipe to stop the flow. Only water will be discharged from the break in the pipe if the operation is completely successful. The lines should be secured in place, all firefighters removed to a safe distance, and the operation continued until the officer-in-charge has been assured by a responsible representative of the company owning the line that the pipe has been shut down and all flammable liquid removed from it.

The above procedure is risky. It is similar to moving in to shut off a valve feeding a flammable liquid fire; however, in this case it is on a much larger scale. If the risk seems too great, it may be necessary to let the fire burn until the source of the spill can be shut off by owners of the pipe line. This is feasible and less risky, providing the spread of the fire can be controlled and the exposures adequately protected.

Oil Tank Fires

Fires in oil tanks present a fearsome spectacle, even to experienced firefighters. They are usually accompanied by dense black smoke intermixed with orange flames. It should be remembered, however, that this heavy black smoke actually offers protection as it screens radiant heat that might otherwise ignite other tanks or buildings in the area.

The two primary problems of oil tank fires is protecting the exposures and extinguishing the fire. Normally the initial lines are laid to protect the exposures. Once water has been played on an adjacent tank to keep it cool, the stream must be applied continuously. Intermittently cooling of an exposed oil tank could cause it to breathe with the potential result of pulling the fire inside the tank.

Most oil fires occur in refineries or in similar types of installations. The first thing the officer-in-charge should do when arriving at the fire is to check with the plant authorities to determine the type of oil burning. Most refineries have their own fire brigades and have predeveloped plans for coping with fires. They also generally have the equipment that will prove most effective for extinguishing the fire. Cooperation between the fire department and plant authorities usually produces effective results.

Generally, the most effective extinguishing agent for oil fires is foam. If possible to do so, it is a good idea to spray water over the burning surface of the oil prior to applying the foam. It is a good practice to spray a small amount of water on the oil and then wait to see what happens. If no slopover occurs, the cooling can continue. A *slopover* is a small spill of oil out of the tank. If there is a slopover, it should be allowed to subside and then more water applied. This can be repeated until the oil is cool. This precooling of the oil will often reduce the amount of foam required for extinguishment and cut down the extinguishing time.

Most large storage tanks have their own permanently installed foam systems. These systems should be used, if possible. However, many times these installations are destroyed by an explosion that precedes the fire. In this case it will be necessary to use portable foam applicators. Most of these applicators are hydraulically raised and equipped with a goose-neck nozzle, which can be placed over the top of the tank and the foam applied. Almost all refineries maintain these applicators as standard equipment. If two applicators are available, the second one should be set up on the opposite side from the first. The foam should be applied for a considerable period of time in order to form a thick blanket over the hot oil. The extinguishment time can be reduced if it is possible to use handlines from a ladder or elevated platform to knock down the hot spots once the fire has subsided.

In some cases an oil tank fire may be encountered where portable applicators are not available. It may be possible in these instances to apply the foam from a ladder pipe or elevated platform. The foam stream should be projected to the far side of the tank and allowed to flow back gently over the surface. If the equipment is available, it might be possible to cut down the extinguishing time by using two ladder pipes or elevated platforms from opposite sides of the tank.

Theoretically, the effectiveness of the foam blanket can be improved by keeping the outside of the tank cool by the use of hose lines while the foam is being applied. The cooling action should reduce the rate of vaporization of the oil, which would have the effect of limiting the amount of vapors available to burn. The overall result would be less heat available to work on breaking down the foam blanket. Regardless of whether or not the theory is true, no harm can be done by keeping the tank cool if the operation can be performed safely while the foam is being applied.

Some oil fires can be extinguished by water alone. Certain viscous oils will mix with water and form an emulsion that acts as a type of foam. Care should be taken when using this method, for it is possible to cause a slopover. To avoid a slopover, or to reduce it to a minimum, the water should first be applied lightly over a small area of the fire and then shut off to see what happens. Once frothing takes place and appears to be capable of being controlled, the water should be applied over the entire surface until the froth forms.

One of the greatest dangers to firefighters at an oil tank fire is the possibility of a boilover. A *boilover* is the expulsion of the contents of the tank by the expansion of water vapor that has been trapped under the oil and heated by the burning oil. Fortunately, most oils are not capable of boiling over, as they lack the basic characteristics required for this to take place.

The basic requirements for an oil to be capable of boiling over are a high viscosity, a wide range of boiling points, and the presence of water and heat. *Viscosity* is the ability to flow. A liquid with a low viscosity will flow easily. A liquid with a high viscosity flows relatively slowly. If water is trapped and heated below an oil with a low viscosity, the water will merely bubble up through the oil. However, if the oil is sufficiently viscous, the steam formed under the oil will lift the entire body of oil and discharge it from the tank.

Crude oil has all the requirements for a boilover. The higher the viscosity of the crude, the greater the chance of a boilover. A boilover should be anticipated any time a tank of crude has burned for more than a few minutes. A boilover can occur suddenly and without warning and will normally be accompanied by heat and flames that will extend several hundred feet into the air. The best action that can be taken when a boilover is expected is to keep everyone several hundred feet away from the burning tank. It is also wise to ensure that all avenues of escape are kept open.

Two other methods of extinguishing oil tank fires are the subsurface injection of air and the subsurface injection of foam. The subsurface injection of air is referred to as the *air agitation method.* The air injected at the bottom of the tank forces cool oil from the bottom of the tank to the surface where it replaces hot oil. This has the effect of limiting the vaporization of the oil by reducing the oil's surface temperature.

Similar results are obtained by the *subsurface injection of foam.* The foam and the cool oil both rise to the surface. A foam blanket is eventually built over the surface to obtain complete extinguishment.

Both of these methods are normally applied by oil company personnel without the participation of fire department members.

FLAMMABLE GAS EMERGENCIES

The two flammable gases most commonly encountered by fire departments in emergency operations are natural gas and liquified petroleum gases (LPG). The most commonly encountered LPGs are butane and propane or a mixture of the two. Both natural gas and LPG are referred to as fuel gases. A *fuel gas* is a flammable gas customarily used for burning with air to produce heat. Natural gas is normally stored and transported as a gas, whereas LGP is generally stored and transported as a liquid. One of the primary differences between the two is in their vapor density.

Vapor density may be defined as the weight of a volume of vapor as compared with an equal volume of air. Vapor densities of less than 1.0 are lighter than air; those greater than 1.0 are heavier than air. This means that gases having a vapor density of less than 1.0 will rise when they escape from their container, whereas those having a vapor density of more than 1.0 will travel toward the ground. Natural gas has a vapor density of less than 1.0. Consequently, it will tend to rise when it escapes from its container. LGP, on the other hand, is heavier than air, having a vapor density of 1.5 or more. This means that escaping gas will travel to and hug the ground while it searches for a source of ignition. Both natural gas and LPGs are odorless; however, both are mixed with a material that gives them a distinctive odor that can be detected well below their explosive limits.

Firefighters respond to two basic types of emergencies involving flammable gases. One type is a leak of the gas and no fire. The other is where fire is involved. The latter primarily applies to fires in transportation lines and storage equipment. For the purpose of firefighting, tank trucks and railroad tank cars are considered as storage equipment.

Nonfire emergencies involving LPG are much more serious than those involving natural gas; however, the procedure for handling the two is basically the same. The primary objective is to keep the escaping gas out of contact with people, out of structures, and particularly out of contact with an ignition source. These three activities are coordinated with the attempt to shut off the source of the leak. The most commonly accepted method used to accomplish the contact objectives is to direct, dispense, and dilute the gas by the use of spray streams.

The extent of the critical area is normally much larger for LPG leakage than it is for natural gas. Natural gas is lighter than air and will tend to travel upward when it escapes from its container. Consequently, it will normally not travel too great a distance in search of an ignition source. However, LPG is heavier than air and will travel greater distances while physically maintaining an explosive area. When it escapes, it will produce a heavy fog-like vapor cloud that assists in identifying the problem area. It is important to remember that the ignitible area extends well beyond the limit of the vapor cloud. In fact, it is good practice to consider that ignitible gases may extend a distance of up to 200 feet beyond the vapor cloud. This potential should be taken into consideration when eliminating ignition sources. (See Figure 10.4.)

The primary danger area for potential ignition is normally downwind from the source of the leak: however, the ignition potential exists on all sides. Initial

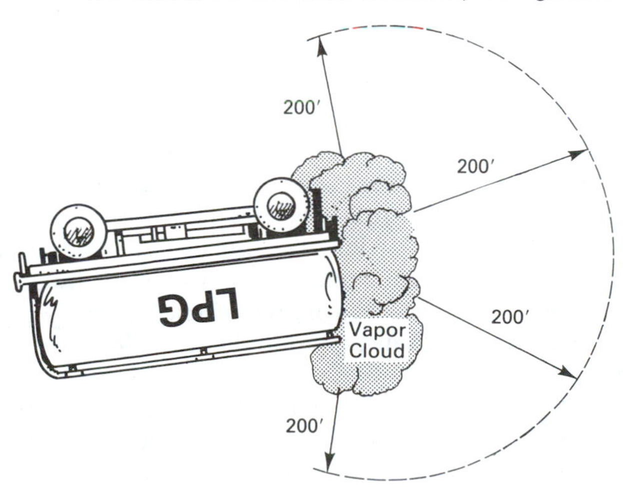

Figure 10.4 Consider that ignitible gases may extend out up to 200 feet

action includes removing all people and shutting off all flame sources such as pilot lights, oil burners, and so on. Other potential ignition sources such as operating engines and electrical motors should also be eliminated. It should be kept in mind that a running engine on a fire apparatus can also be a hazard. It is possible for a diesel engine to "runaway" due to a mixture of LPG and air or gas and air being taken into the engine and compressed rather than the engine compressing nothing but air. If the spill is widespread, it may be wise to shut down all electrical power in the area.

There are a couple of important points that should be kept in mind while directing, dispersing, and diluting the gas, removing the people, and eliminating the sources of ignition. LPG is much more hazardous and will spread out over a much larger area than is normally thought. For example, a container holding propane will produce approximately 270 cubic feet of gas for every cubic foot of liquid. If the gas finds an ignition source, the result will be a very violent explosion, with the fire involving the entire vapor area all the way back to the source of the leak. This could be a short distance or it might extend over several thousand feet. Consequently, it is good practice to operate always as if ignition could occur and in a manner that the resultant damage will be minimized if it should happen. Lines should be laid that will be needed to protect the exposures, and personnel should not be placed in jeopardy while performing their duties.

In many cases the leak can be stopped by closing a valve. However, in a large number of situations the valve needing closing will be in the vapor area. Personnel should never be sent into the vapor area to shut off a valve unless a sufficient number of protective streams are provided to ensure maximum protection in the event that ignition occurs while they are performing the operation. The personnel assigned this task should always wear full protective gear, including breathing apparatus.

One last caution regarding LPG spills: It is good practice to approach the reported spill from the windward side whenever possible. Apparatus should be parked in the center of the street, a safe distance from the reported spill. Parking in the center of the street assists in eliminating the apparatus as a source of ignition, as the spilled LPG will probably travel and remain in the gutter area if the leak is not too extensive.

Fire Incidents

A basic principle of fighting fires in combustible gases is *never extinguish the fire unless the source of the leak can be stopped.* (See Figure 10.5.) Although there are situations that will require deviation from this principle, it is good practice always to consider the principle prior to taking other action. The principle is based on the concept that extinguishing the fire prior to eliminating the leak will normally create a much larger hazard than the fire itself. The escaping flammable gas will spread over a wide area, with the potential result of an explosion that could do extensive damage and possibly cost a number of lives.

Small natural gas fires in buildings can generally be extinguished by shutting off the nearest valve. If a fire occurs at the meter, it will be necessary to shut off the valve at the street, as the valve at the meter controls only the flow of gas on the

Figure 10.5 Never extinguish the fire unless the leak can be stopped

discharge side of the meter. If a fire occurs in the street as a result of a break in the main transmission line, the general practice is to protect the exposures and let it burn until flow to the line has been shut off by gas company personnel.

Minor fires at the relief valve of small LPG storage tanks can be extinguished by cooling the tank with water. This will relieve the built-up pressure within the tank that caused the relief valve to open and the valve will close.

The most challenging flammable gas fires occur in LPG transportation tank trucks. The basic tactics of handling these fires apply to railroad tank cars and permanently mounted storage tanks; consequently, the discussion of firefighting operations will be limited to tank trucks.

The initial action required when arriving at a large fire involving a LPG tank truck is to protect the exposures and cool the burning tanker. This will probably require large quantities of water and a number of hose streams. If sufficient water is available, it may be best to set up heavy stream appliances. Portable heavy streams appliances not only provide large quantities of water but can be left in place to continue the cooling action if it becomes necessary to evacuate the area suddenly.

Keep in mind that lines should not be set up at the ends of the tank, as the tank can become a flaming rocket if one of the ends lets go.

If flame is impinging on the burning tank, it is imperative that water in large quantities be applied to the impinged areas as quickly as possible. The overall objective is to prevent the tank from "bleving." A *BLEVE* (boiling-liquid, expanding-vapor explosion) is a devastating effect caused by failure of the container. It is the result of impinging flames building up pressure within the tank faster than it can be relieved by the spring-loaded relief valve. The buildup of pressure is even more positive when the tanker is turned over. When the tanker is upright, the relief valve is in the vapor area. When the tanker turns over, the relief valve is in the liquid area and as a result is unable to relieve any vapor pressure buildup.

The BLEVE is generally a result of the combination of the pressure buildup and the lost of strength of the metal caused by the heat of the flame impingement. Failure of the metal will usually occur at the point of flame impingement. The buildup of pressure will cause a longitudinal tear in the metal, with the resultant release of large quantities of the pressurized LPG. Because the burning fuel is an immediate source of ignition, the result is a tremendous explosion accompanied by a huge fireball. The danger area for the radiated heat from the fireball and the concussion from the explosion could be as much as half a mile. People have been killed as far away as 800 feet as a result of flying fragmented metal. Anyone doubting the power of a BLEVE should take the time to research an incident that occurred in Kingman, Arizona on July 5, 1973. In this incident, 15 people were killed, 13 of them firefighters.

There are a couple of factors regarding BLEVEs that should be kept in mind when planning operations. Most BLEVEs occur when the containers are ½ to ¾ full of liquid. A tanker that is nearly full when the fire is initiated will eventually contain liquid within these proportions if a leak in the tank is feeding the fire. Records indicate that the time between the initiation of the fire and the BLEVE varies from a few minutes for small containers to a few hours for very large containers. The best preventive measure against a BLEVE is the playing of large quantities of water on the exposed tank. Sufficient cooling will keep pressure from building up within the tank and prevent the weakening of the metal. It is good practice to set up master streams for keeping the tanker cool, and to withdraw personnel to a safe distance. This is particularly true if flames have been impinging on the tanker for any length of time when the firefighting forces first arrive. Although a BLEVE should always be anticipated, there are two positive warning signs that should not be ignored. One is an increase in the pitch of the sound made by the escaping gas. This is generally a result of a rapid increase in the rate of vented gases. The other sign is a discoloration of the tank near the flame impingement area.

If indications are that the fuel feeding the fire can be stopped by closing a valve, and it is reasonably safe to attempt it, then every effort should be made to do so. The same general precautions taken for shutting off a leaking source when there is no fire should be observed; however, where fire is involved, the number of lines used for protection of personnel will normally need to be increased.

Once the flow has been stopped, plans can be made to extinguish the fire. The best agent for fire extinguishment is dry chemicals providing there is a sufficient amount available to do the job. The initial discharge of the dry chemicals can be started close to the leak or from some distance back. Several discharge lines should be used if the decision is to make a close approach. All nozzles should be directed at the source of the leak and opened simultaneously. This will normally knock down all the fire close to the source of the leak. Additional dry chemicals can be used to extinguish any remaining fire.

With the second approach, the dry chemical nozzles are opened up some distance from the fire. The firefighters manning the nozzles should be spaced across in a line. The objective is for the nozzlemen to work in unison to produce a wall of dry chemicals as they advance on the fire. The nozzles should be operated up and down in a vertical alignment. This type of approach not only has the effect of quickly knocking down the fire but also provides a heat shield for the attacking nozzlemen.

Occasionally fire departments will encounter LPG tanker fires in areas lacking sufficient water to fight the fire effectively. This might occur in rural areas or perhaps on freeways where the incident is remote from a good water source. In these cases it is usually wise to start an immediate evacuation of the area. If time permits, unmanned heavy stream appliances can be set up to use whatever water is available. It is possible that this action could prevent a BLEVE. The overall plan, however, should be one where a BLEVE is anticipated. Operations should be directed to minimize the loss if the BLEVE does occur. It is good practice to evacuate the area at least ¾ of a mile, and preferably 1 mile, around the fire. (See Figure 10.6.)

ELECTRICAL FIRES

For purpose of definition, an *electrical fire* is a fire in energized electrical equipment. Once the circuit has been deenergized in electrical equipment, the fire becomes a Class A fire or a Class B fire, depending on the material burning. It can then be handled in accordance with the material involved.

The primary problem for firefighters when fighting electrical fires is the potential for electrical shock. Consequently, whenever possible, the electrical circuit to the involved equipment should be deenergized prior to an attack being made for the purpose of extinguishing a fire.

Occasionally it is necessary to extinguish an electrical fire. Most electrical fires encountered by fire departments are small and can be extinguished by the use of a portable extinguisher. Extinguishers classified for use on electrical fires contain a nonconductive material. Carbon dioxide, dry chemicals, and halon all meet this requirement.

Dry chemical extinguishers are extremely effective for use on electrical fires but generally require some clean up after the fire has been extinguished. The multipurpose extinguisher also presents a potential corrosion problem if all the powder is not cleaned up soon after the fire has been put out. Consequently, it is best not to use it where the powder may get onto metal parts that are difficult to reach.

Carbon dioxide and halon are noncorrosive and leave no residue. One of these extinguishers should be given first choice when the fire involves sensitive electrical

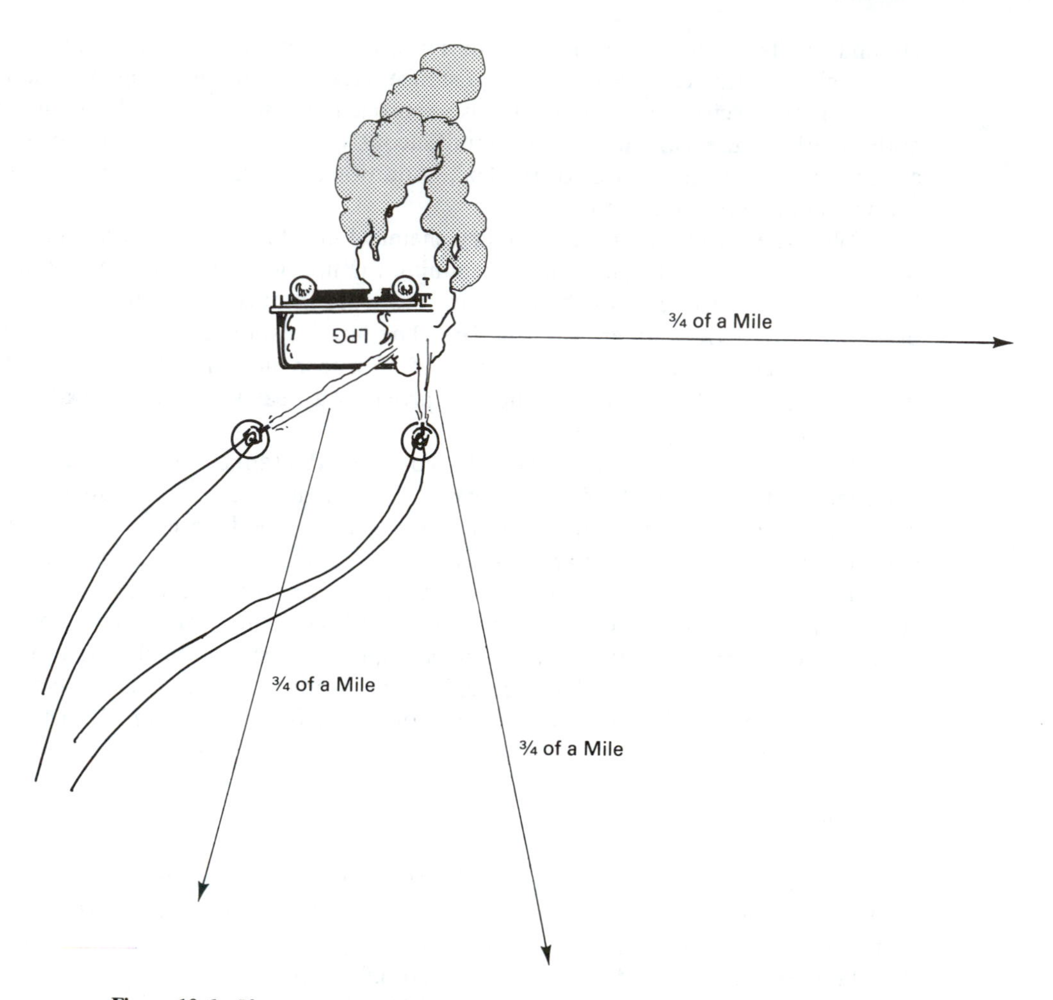

Figure 10.6 If necessary to abandon the fire, evacuate the area for at least 3/4 of a mile

equipment. The use of dry chemicals on fires in sensitive electrical equipment (such as computers) could result in more damage than that produced by the fire itself.

In some cases it may be necessary to use water on deenergized electrical equipment. A good example is a fire in an electrical motor. Water can be used after the circuit has been shut off. The circuit can generally be deenergized by turning off a switch or pulling a plug. Electrical motors get very hot and it may be necessary to apply water for some time to cool it off. The water can do no additional damage since the windings on the motor will have to be rewound in any event.

The electrical wiring in many areas of some cities is run underground. Underground fires in manholes and street transformers are not uncommon where these systems are found. These fires present special problems to firefighters.

It is good practice when responding to a reported fire in underground equipment not to park the apparatus over a manhole cover, regardless of whether or not smoke is issuing from the opening. Fires in these systems have a habit of collecting flammable vapors and blowing a manhole cover high in the air when ignition takes place. As a precaution measure against this potential, all people within a block of a smoking manhole or vault should be kept back a safe distance from all manholes. After a call for help from the electrical department has been placed, all main service switches and fuses to all buildings within the block of the emergency should be pulled. It is a good idea not to use the bare hand to pull the switch, as the switch may be charged. If possible, a pike pole or axe can be used for this purpose. Rubber gloves provide good protection when removing fuses. Disconnecting the power provides some, but not positive, protection against the problem extending into buildings. As a precautionary measure against further extension into buildings, a constant check should be made of the electrical entrances into buildings until the officer-in-charge has been informed by a responsible person from the electrical department that the situation is under control

Unless it is absolutely necessary for the purpose of making a rescue, underground vaults and manholes should not be entered. If entrance is necessary, firefighters should wear breathing apparatus and carry explosive-proof flashlights.

Water or other conducting liquids should never be poured down transformer vents. No harm can be done by discharging carbon dioxide into the vents but there is no assurance that it will do any good. The best procedure is to keep everyone a safe distance away and not do anything unless requested by representatives from the electrical department.

Use of Water

Occasionally fires will occur in electrical equipment that cannot be extinguished by portable extinguishers. In these instances some departments prefer to let the fire burn and protect the exposures until all power to the electrical equipment can be shut off. Other departments feel that water can be used effectively as long as proper precautions are taken.

The first precaution is *never use contaminated or salt water.* The conductivity of these materials provides too high a risk. The second is *avoid the use of straight streams.* In fact, if they are available, it is best to use nonadjustable fog nozzles. The next best bet is spray nozzles. If spray nozzles are used they should be adjusted to provide the smallest droplets possible prior to advancing on the fire and not changed during the extinguishing operations.

Nozzlemen directing water on an electrical fire should stand on a dry spot. Standing in water will provide a good ground in the event of electrical flow down the stream. Contrary to some thoughts, boots are not a good insulator. Some protection is afforded if the nozzleman wears rubber gloves. (See Figure 10.7.)

If all safety precautions are observed, it is safe to use water in fog form from any distance where the electrical equipment is of low voltage. Low voltage is that under 600 volts. Additional precautions, however, should be taken when using water on high-voltage equipment.

High-voltage equipment is normally identified by signs. If there is any doubt as to the voltage, it is best to assume the equipment is high voltage.

Many tests have been conducted on the safe distance required when water is

Figure 10.7 Water can be used if proper precautions are taken

used on high-voltage systems. Results vary, making it difficult to establish any hard and fast rules. The tests do indicate that the smaller the nozzle tip or its equivalent, the less chance for voltage flow. Past practice indicates that it is relatively safe to use water on high-voltage installations as long as the nozzleman stands on a dry spot and the water is discharged on the fire at the maximum reach of the stream.

BRUSH FIRES

Statistically, brush fires have played a prominent role when evaluating those fires that have caused losses of $1 million or more. They have also been responsible for the death of a number of firefighters. Although this chapter discusses brush fires, the tactics involved and the safety precautions required are applicable to similar types of ground cover fires that occur throughout the country.

Brush fires can occur during any month of the year; however, the most critical period is one following an extensive number of days of low humidity. *Relative humidity* is the amount of moisture in the air compared to the amount the air can hold at that temperature. The lower the humidity, the drier the air. Low humidity allows the brush to become timber-dry, making it extremely susceptible to any source of ignition. Some of the common sources of ignition are carelessness, an act of nature, or a deliberate application of flame. The overall result is the same. If the fire is reported in its infancy and properly handled, the loss will be minor. If the report is received late or the fire is improperly handled, it can quickly develop into a major conflagration.

Factors to Consider

There are a number of factors that should be considered when planning an attack on a brush fire. Each of these should be carefully evaluated, as they will have a direct bearing on where the fire should be attacked and what method of attack will prove to be most successful.

Wind and draft conditions. This is the first and most important factor to be considered in fighting brush fires. The fire will act completely differently when the air is still than it will when a stiff breeze is blowing. The wind has a dual adverse affect on the fire. It causes the fire to move rapidly and it also brings in fresh oxygen to feed the fire. Not only should the wind be evaluated when deciding how to make the initial attack but it should be watched continuously during the entire fire operation. Winds have a habit of rapidly changing direction. This action has trapped firefighters by suddenly blocking their avenue of retreat. Everyone on the fire line should continuously be on the alert for flames that suddenly stand straight up. This is a strong indication that the wind direction is about to shift. (See Figure 10.8.)

Canyons have a definite and varying affect upon the winds. Each canyon

Figure 10.8 Be careful of a wind shift when the flames stand straight up

should be considered as a flue. Canyons act similarly to vertical openings in a building. The wind can be pulled in one direction in the main canyon and can be flowing in different directions in the lateral canyons. This can result in the fire burning in various directions at the same time. These changing conditions can make it extremely hazardous for firefighters. Personnel should never be sent into a canyon ahead of the fire when the fire is close to a steep slope, as the fire will rush rapidly toward a ridge and likely trap the firefighters.

The ridges. A ridge is the top of the hill, the place where two opposing hill slopes come together. The ridge is generally the safest place to work on the fire and the most likely place of stopping it. When going in on a ridge, it is safer to take the one on the windward side. It should be remembered, however, that the fire will travel rapidly uphill, which makes it almost impossible to stop until it reaches the ridge. The steeper the slope, the faster the fire will travel. There is usually a burst of flame as the fire reaches a ridge due to fresh oxygen being pulled into the fire from the opposite side of the ridge. Because of this action, it is best that hose lines and firefighters be positioned just over the ridge from an advancing fire.

Firebreaks. Firebreaks serve a dual purpose at a brush fire. Most importantly, they may stop the fire. Second, they provide a means of getting into areas that would otherwise be inaccessible.

Firebreaks are of two types. One is referred to as a *permanent break*. In many parts of the country, permanent breaks are constructed by governmental workers in strategic locations along the ridges well in advance of the fire occurrence. A well-designed firebreak system will create a patchwork of hazardous areas and provide protection against fire spread, much as fire walls do in buildings.

The second type is referred to as an *emergency break*. These breaks are made at the time of the fire. They vary in width, depending on the sharpness of the ridge, the brush conditions, and the other factors contributing to the seriousness of the fire. Under average conditions, a break from 10 to 12 feet in width is sufficient. (See Figure 10.9.) If possible, the break should be started on the point of the ridge where it is anticipated that the fire will first reach. The break should cover the top of the ridge and extend down part way on the opposite side from the fire approach. All cut brush should also be thrown down this side.

Available equipment. Both motorized and hand equipment is used on brush fires, but a good portion of the work is done with hand tools. Probably the most useful universal tool is the long-handled, round-point shovel. Its primary use is throwing dirt on the fire, but it can also be used to swat the fire, scrape the fire line, and dig a safety hole if necessary. If things get critical, it can also be used as a face shield on a hot fire.

Some of the cutting tools used for making an emergency firebreak are the brush hook, the axe, and the Pulaski tool. The popularity of each of these tools varies in different parts of the country. Care should be taken when using these tools so as not to injure other members of the crew.

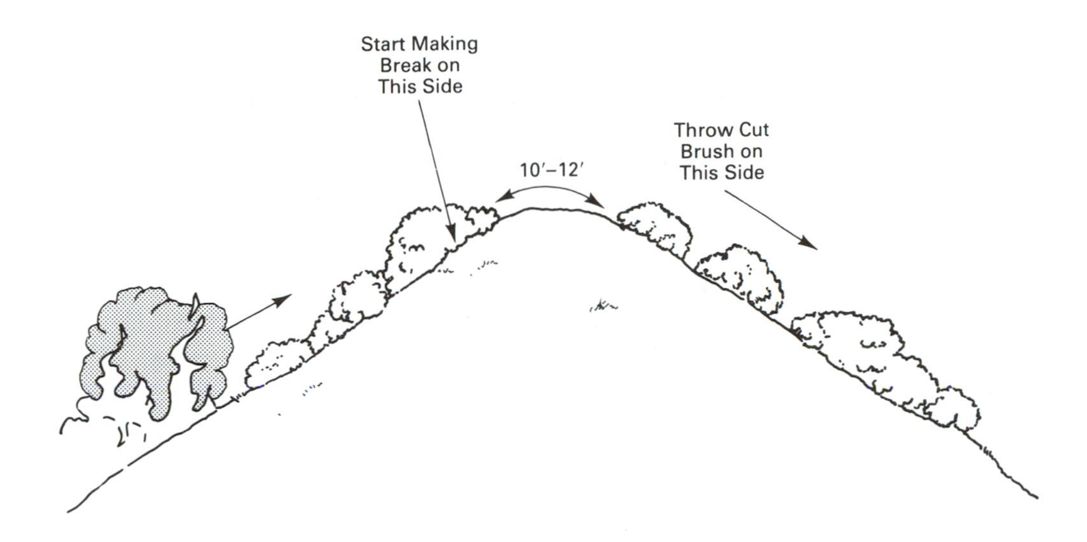

Figure 10.9 Emergency firebreaks should be 10–12 feet in width

Air support. The amount and type of air support is a prime consideration when developing plans for stopping a brush fire. Helicopters are extremely useful for reconnaissance work, dropping smokejumpers into strategic locations, delivering equipment, making rescues and, serving as an elevated command post. Both fixed-wing aircraft and helicopters are useful for water drops. The aerial drops may be regular or viscous water. Although it is very unlikely that water drops from aircraft or helicopters will completely extinguish a fire, they do play an important part in slowing down the fire to allow ground crews to complete the job.

The most effective use of helicopters for making water drops is when a team is established, consisting of three helicopters and a land-based support group. The support group will seek out a location for a helipad that is not too far away from the fire and in close proximity to a good water source. Many of these helipads are determined long in advance of the actual fire. A line is laid from the water source to the helipad, using large diameter hose. The helicopters land at the helipad in a planned coordinated operation. The ideal is for one chopper to be dropping water at the head of the fire, the second about to land to pick up a fresh supply of water, and the third just taking off and heading in to make its drop on the fire.

Exposed structures. Any structure in the path of a brush fire should be considered as an exposure. When evaluating the potential path of the fire travel, constant consideration should be given to the possibility of the direction of travel changing due to wind or terrain conditions. In those areas where fire apparatus can be safely parked and firefighters work safely, companies should be dispatched to protect the exposures and lines should be laid long before the arrival of the fire. Any movable burnable objects outside of the house, such as lawn furniture, should be moved into the garage or house. All windows and drapes in the house should be closed. After all outside preparations have been made and the brush in front of the house thoroughly wetted, the line should be placed in a safe location and the crew members should go inside the house and wait for the approaching fire. After the fire has swept by, the firefighters can go outside and knock down any fires that may have started.

There is a strong tendency on the part of some officers and firefighters who are not familiar with brush fires to believe that it is possible to stop the advancing fire by the use of water. All those involved in brush firefighting should always remember the strength of an advancing brush fire. Water is effective for putting out a fire, but this does not mean that a loaded 2½-inch line is capable of stopping a fire that is moving rapidly up the side of a hill. The fire will preheat the brush and other combustible material well in advance of the actual flames. A basic principle of fighting brush fires is *never go down through unburned brush to meet a fire burning uphill.* This rule applies both in the protection of exposures and during actual fire extinguishing operations. It is wise never to underestimate the strength nor the speed with which a brush fire can move. Firefighters have been killed because they were either careless or not aware of the potential danger.

Brush conditions. The density of the brush growth should always be considered when planning firefighting operations. This, together with the wind and draft conditions, will have a definite affect on how the fire is to be fought. Under favorable conditions, a direct attack may be practical in light and medium brush; however, if the brush is heavy or the winds adverse, other means of attack will probably have to be made.

Fire Extinguishment

The initial response to a reported brush fire will vary according to the jurisdiction in which it occurs. Some departments include the movement of tractors and helicopters, whereas others dispatch only engine companies. The first arriving officer should be familiar with the number and types of companies dispatched on the first alarm. This will allow the officer to determine what is at his or her immediate disposal and what additional help may be required.

Many departments use engine companies that have tank wagons as part of the routine dispatch to brush fires. A tank wagon normally carries 700 or more gallons of water and a good supply of small lines. Small brush fires can generally be handled with two or three of these units. If the fire is of any size whatsoever, additional help will probably be required. The first-in officer should evaluate all conditions and then call for the help that is necessary, including a sufficient number of units to allow the officer to "fail safe." The officer should then place his or her developed plan into operation.

Fire extinguishment of larger brush fires is normally made using a combination of the direct and indirect attack. The direct attack is one where action is taken directly against the flames of the fire. It may be made by using hose lines, hand tools, or a combination of the two.

The indirect attack consists of applying control techniques some distance ahead of the fire to stop its progress. It generally consists of creating an area ahead of the fire through which the fire cannot burn. This can be done by using a natural barrier such as an old burn, a permanent firebreak, creating an emergency firebreak, or wetting down the brush. Usually a combination of these is employed.

After calling for the required equipment and personnel, the first-in officer should request that any air support available make a quick survey of the area ahead of the fire's progress to determine if there are any structures that may have to be protected. The aircraft should then be directed to make a water drop on the head of the fire. Instructions should be given to incoming tractor units as to where to start making an emergency firebreak. Sometimes their most useful work can be done on a flank, working on the fire's edge. If the brush conditions are such that it is possible to do so, the first-in officer should then have the crew lay small lines to the hot side of the fire. The second-in company officer should lay lines to the opposite flank. Both officers should carry portable radios to provide directions for other units and to relay information regarding the fire's progress to the chief officer, who should arrive within a short period of time.

Both companies should continue to advance their lines along the flanks of the fire in an effort to contain the main body of the fire while it moves toward the projected areas where it can be stopped. Additional arriving units can lay lines to support them in their progress.

It may be possible that water is not available or that the fire is beyond the reach of hose lines. In these cases it will be necessary to make the direct attack by using hand tools. The objective will be to separate the fire from the fuel by cutting a fire line along the edge of the fire. Dirt can be used to extinguish the fire during this process.

Shoveling dirt on a fire is hard, back-breaking work. It is important that every shovel full of dirt extinguish as much fire as possible. The most effective method is to get a good shovel full of dirt and throw it with a sweeping motion. This will generally kill several feet of fire. If the shovel is used for clearing brush, it should be done in a manner that the brush will not have to be handled more than once.

Water drops from aircraft can be extremely effective for extinguishing or controlling spot fires. If air support is available, there should be no hesitation in calling for it.

It is possible for sparks to be blown in advance of the main fire to start spot fires. A constant lookout should be maintained for these spot fires, for they can rapidly develop into larger and more dangerous fires. Consequently, they should not be neglected. Action should be taken as quickly as possible to extinguish them. However, crews should not be put into a precarious position in order to do so. It is good practice to have roving units available to be on the lookout for spot fires and to handle them where possible.

The ridge is the safest and best place to stop the fire. To stop the fire at the ridge requires that sufficient personnel get to the ridge ahead of the fire. Crew members should be stationed 40 to 50 feet down the slope of the ridge on the opposite side from which the fire is traveling. With the exception of spot fires that start on their side of the ridge, which could place them in jeopardy, crew members should normally be safe at this location. Any spot fires that develop should be handled as quickly as possible. Failure to do so is likely to result in the firefighters being trapped between two fires.

Two conditions must exist for the fire to be considered under control. One is that the main body of the fire reaches the prepared breaks and diminishes in size. The other is that the spread at the flanks has been cut off. Cold trailing and patrolling should be started once the fire is considered to be under control.

Cold trailing is the process of making a clear break between the burned out area and the unburned brush while extinguishing any burning embers. The width of the break should be at least as wide as the brush is high and preferably wider. It is the only positive means of ensuring that the fire is out. It is done through the combination use of hand tools and mechanized equipment. The area should be raked clean of leaf mold and other combustible material. All material should be thrown back into the burned area. If water is available, it can be used effectively as an aid in cold trailing. All stumps or other burning material must be completely extinguished. Care must be taken when water is used to ensure that sparks or embers are not driven into the unburned brush. (See Figure 10.10.)

Patrolling consists of maintaining a vigilant watch over the entire area to ensure that the fire does not rekindle. Extra precaution should be taken at daybreak, as this is the period when the danger of rekindle is the greatest. The danger is caused by a change in the wind conditions. The wind normally dies down during the night and starts picking up at daybreak.

Occasionally it may be necessary to backfire as a portion of the indirect method of attack. *Backfiring* is the intentional setting of a fire to establish a defense perimeter for the fire or as an emergency safeguard for personnel. Backfiring may become necessary due to extra heavy brush, adverse wind conditions, or on those occasions when it is not possible or practical to get ahead of the fire and establish a firebreak. Backfiring is a risky business and should be done only as a last resort.

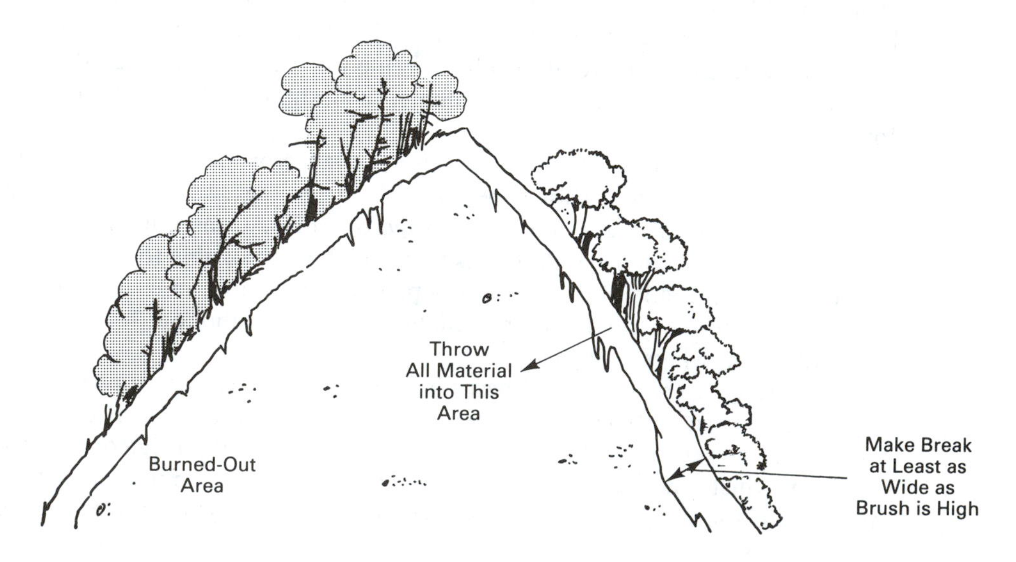

Figure 10.10 A cold trail break

Except in extreme emergencies, backfiring should not be done without the direct approval of the officer-in-charge or the person having the authority to issue this order.

Except in an emergency where personnel are in danger, it is good practice not to start a backfire unless control lines are in existence around the entire area to be consumed by the backfire and all people and animals have been removed from the area. The backfire should be commenced at the nearest safe natural or manmade firebreak of sufficient width that will permit the backfire to burn against the wind. The backfires should be started in a wide line with the ends secured at a safe break whenever possible. It is good practice always to have a contingency plan in the event the backfire procedure fails. However, if the winds hold steady and the backfiring is done properly, it should prove to be a successful operation.

PERSONNEL CONSIDERATIONS

1. Occasionally firefighters are trapped at a brush fire. It may be possible that the only way to save themselves is by starting a backfire. If this is the case, there should be no hesitation to do so.
2. Every person going into a brush fire should be equipped with a canteen. This point cannot be overstressed. It is good practice, however, not to drink too much water when overheated or exhausted. This practice could cause severe nausea.
3. Operating in the burned area is safer than operating in an unburned area.
4. Personnel should remain calm if caught in a heavy smoke situation. It should be remembered that the freshest air is close to the ground. Good air is also normally trapped underneath a firefighter's clothing. It is important not to overlook this fact in a critical situation.
5. If necessary to retreat from a fire, it is best to work toward the flank and to the rear of the fire front.
6. If personnel are trapped on an apparatus ahead of an advancing fire, it is best to climb under the apparatus and remain there until the fire has swept by. *Do not* take a line off the apparatus and try to protect the apparatus from the advancing fire. Firefighters have been killed trying to do so.
7. Firefighters should be brought off the fire line, fed, and allowed to rest at appropriate intervals. Those replacing them should be thoroughly briefed on what has been done and what remains to be done. It is best that reliefs be made prior to nightfall. Those working on the line at night should, at a minimum, be equipped with warm clothing, a good flashlight, a canteen, and the necessary tools to complete the assigned tasks.
8. If available, fire shelters should be carried by all personnel trained in their use.

APPARATUS CONSIDERATIONS

1. If it becomes necessary to park an apparatus and leave it, the apparatus should be parked so as not to block other apparatus and be pointed in a direction of safe retreat.

2. An apparatus should not be driven through heavy smoke unless there is positive knowledge of where the pathway leads.
3. It is well to remember that it is possible for the engine of an apparatus to stop if engulfed in heavy smoke. The engine of the apparatus operates by combining air with fuel. No air, no ignition. This point is particularly important to remember if lines are being supplied from a pump on the apparatus.
4. The headlights should be on and the windows closed when traveling through smoke. It is also good practice to sound the horn intermittently.
5. A line should be laid to protect the apparatus whenever the apparatus is being operated at a brush fire.

PIER AND WHARF FIRES

It is not unusual to find an experienced firefighter who has recently been transferred to a harbor battalion to start talking about dock fires. One of the first things this individual learns comes as a shock—there is no such thing as a dock fire. Much to the surprise of many people, a *dock* is an open body of water in which a vessel floats when tied up to a wharf or pier. A *pier* is a structure that projects out into navigable waters so that vessels may be moored alongside of it. It is usually longer than it is wide. A *wharf,* on the other hand, is a similar type structure that is built along and parallel to navigable waters. For the purpose of firefighting, piers and wharfs may be considered the same. In this section of the chapter, both will be referred to as *wharfs.*

There is no universal standard for the construction of wharfs. Some are built of fire-resistive construction, others are constructed of wood. Those constructed of wood are the ones that present a problem to firefighters. Although the construction of wharfs will vary in different parts of the country, an explanation of the construction features of those found in some parts of Los Angeles harbor will provide a general idea of the problems involved.

For the most part, the wharfs are built on pilings that have been driven into the sand by pile drivers. The diameter of a piling is approximately that of a telephone pole. The pilings are capped by 10″ by 14″ stringers. The deck is constructed of 4″ by 14″ planking laid over 4″ by 14″ joists. Sometimes the deck is double thickness; at other times it is paved with a 3-inch coating of asphalt. On the land side of the wharf is a concrete bulkhead, which prevents access to the underside of the wharf from the land side. Large pipes are often attached to the underside of the joists, which provides a barrier to the effective use of streams from the water side.

Access to the underside of the wharf can be made by climbing down a permanent wood ladder that is found on the water side of the wharf. These ladders are strategically placed approximately every 100 feet. The ladders extend from the top of the wharf to fender logs that are chained to the pilings at water level. Part way down the ladder are stringers and bracing, which can be walked on to reach various parts of the underside of the wharf. The pilings are heavily coated with creosote before being driven into the sand and most of them are covered with a heavy oil that has accumulated over the years. The oil on the pilings can be ignited very easily and will give off a heavy, acrid smoke when burning.

The first arriving company at a reported wharf fire is safe in driving directly out onto the wharf. If a small amount of smoke is rising from under the wharf, it is generally possible to take a small line to the underside and extinguish the fire. If a permanently mounted ladder is not in the vicinity of the smoke, a roof ladder may be used effectively to reach the underside. If complete extinguishment is not possible, the line can be used to control the fire while planking is pried up. Small lines from fireboats can also be used effectively for this purpose. If scuba (Self-Contained Underwater Breathing Apparatus) members are available, it is good practice to have them in the water any time a member of a land company is working under the wharf. This measure is a safety factor in the event a member of a land company falls into the water.

A large amount of smoke rising from under the wharf indicates the potential for a rapid spread of the fire. In this case it is still safe to drive the apparatus out onto the wharf, as collapse of the wharf is unlikely for a long period of time. A sufficient amount of hose should be removed from the apparatus and a lay made back to a hydrant on the land side of the wharf. It is important that the apparatus be removed from the wharf because the wharf construction and the prevailing winds could rapidly spread the fire in a horizontal direction with the wind and trap the apparatus on the wharf. If it appears that the fire may become severe, it is important that boats, ships, or other mobile equipment be moved as quickly as possible. This is important not only for the purpose of protecting these units but also to provide access to the fire for the responding fireboats.

The overall strategy when there is a strong potential for the fire spreading is to get ahead of the fire on both sides and stop its advance by setting up water curtains while attempting to extinguish the fire. Although it is important to try to set up water curtains on both sides of the fire at the same time, priority should be given to the leeward side if a decision between the two must be made.

The tactics used to set up the water curtain will depend on whether or not the department has a well trained scuba team. The scuba team should be given priority if one is available, as they are able to set up the protective curtain quicker and more effectively than can be done by the alternative. Most scuba teams are now equipped with floats on which 1½-inch spray nozzles have been permanently attached. These floats can be interconnected and floated under the wharf and tied into place. The nozzles can be adjusted by team members to ensure that a solid protective screen is being provided. Once this equipment is in place, the frogmen can be removed and used to make a direct attack on the fire. (See Figure 10.11.)

If a scuba team is not available, it will be necessary to take a position well in advance of the fire spread and pry up planking or cut holes in the decking to set up distributor nozzles to create the water curtain. A number of distributors will be needed to provide an effective stop. The distributors should be set up perpendicular to the wharf's edge. The first should be placed near the land side of the wharf and others located a maximum of 10 feet apart until the wharf's edge is reached. The last one should be placed over the edge of the wharf to prevent the fire from getting around the outside of the curtain.

Figure 10.11 An effective water curtain can be set up by scuba members

This method is time consuming and is usually not successful if the fire moves fast. In fact, if the wharf is paved, it is almost impossible to get a water curtain in place in the time available.

Some wharfs have manholes or openings permanently installed that may be used for the placement of distributors. These are not normally located exactly where needed but might serve some useful purpose while the attempt is being made to cut through the wharf's decking. Regardless of whether holes are cut or the permanent installed holes are used, it is important that small lines be laid to protect those working prior to opening the deck. It is good practice that breathing apparatus also be worn.

A direct attack on the fire should be made while the water curtains are being put in place and continued until the fire is extinguished. The first attack is generally made by a fireboat approaching from the leeward side and running parallel to the wharf while sweeping the underside with a battery of heavy streams. Firefighters on the attacking boat should wear breathing apparatus, as an approach from the leeward side places them directly in the path of the heavy smoke being given off by the fire. Unfortunately, due to the mass of pilings, timbers, piping, and so on, it is very unlikely that the fire will be extinguished on the original sweep. It can be expected that some of the fire will remain unaccessible to the fireboat's streams regardless of how many sweeps are made. This is due not only to the hidden areas created by the construction but also because the fireboat's streams cannot be projected from a low enough location to reach all parts of the underside.

Similar to a structure fire where it is necessary to go inside to completely extinguish the fire, it is necessary to go under the wharf to get at all of the fire. The best method of doing this is to put members of the scuba team in the water and send them under the wharf. Most scuba teams are equipped with floats that can be attached to the hose lines to keep them from sinking while the advancement is being made. Scuba members working under a burning wharf are probably in a safer working environment than a hoseman working a line inside a burning building. If the wharf begins to collapse or if fire spreads over the water due to a flammable liquid spill, a scuba member merely has to roll over, dive down 10 or 15 feet, and swim to a place of safety.

If a scuba team is not available, it may be possible to use a small vessel such as a rowboat to take lines under the wharf to complete the extinguishment.

The primary task of land-based companies, other than trying to establish a water curtain, is to protect the exposures. The exposures include any cargo on the wharf, the structure located above the wharf, and possibly adjacent wharfs. If possible to do so, the best method of protecting any cargo on the wharf is to have it moved to a safe location.

Overhauling a wharf fire may become a difficult task. In addition to getting lines under the wharf to thoroughly wet it down, it may be necessary to ply up planking to get at hidden spots. It is generally necessary that heavy equipment be used. Jackhammers can be employed if it becomes necessary to remove paving.

REVIEW QUESTIONS

FLAMMABLE LIQUID FIRES

1. From a technical standpoint, what is a flammable liquid?
2. What is meant by the term *miscible?*
3. What are hydrocarbons?
4. What are polar solvents?
5. What are two examples of flammable liquids that are a mixture of hydrocarbons and polar solvents?
6. What are some of the precautions that should be taken when approaching a flammable liquid spill?
7. What determines the size of a flammable liquid fire?
8. How might a small flammable liquid fire that is not being fed from a continuous source be extinguished?
9. Which portable extinguisher should generally be given first choice for extinguishing a flammable liquid spill fire?
10. What material should be used for extinguishing large flammable liquid spill fires?

11. How should the attack on a flammable liquid spill fire using foam be made?
12. What is the possible dual effect of using water to extinguish flammable liquid fires?
13. What is one of the better methods of attacking a flammable liquid spill fire using water?
14. If only water is available to attack a flammable liquid spill fire, when is it a good idea not to use it?
15. What are the factors that will determine the magnitude of the problem when a flammable liquid spill fire is being fed from a continuous supply source?
16. What four things must be done to solve the problem when a flammable liquid spill fire is being fed from a continuous supply source?
17. What attack method should be used to shut off a valve that is supplying a flammable liquid spill fire?
18. What are some of the methods used to stop the spread of a flammable liquid spill fire?
19. What are the primary problems when a flammable liquid tank fire is located in a diked area?
20. What might be used to extinguish the fire in a small tank of hydrocarbon fuel?
21. What material should be used to extinguish a fire in a large tank of hydrocarbon fuel?
22. What is generally the weakest part of a flammable liquid vertical storage tank?
23. What action should normally be taken to protect exposed flammable liquid storage tanks when the tank on fire is to be left to burn itself out?
24. What is the weakest part of a vertical flammable liquid storage tank?
25. What should be expected if an end of an exposed horizontal tank lets go?
26. What is about the minimum amount of water required to protect an exposed side of a horizontal flammable liquid storage tank?
27. What is one of the basic principles that should never be ignored when playing water on an exposed horizontal tank?
28. If it becomes necessary to evacuate an area because there is an insufficient amount of water to keep a horizontal tank cool, how far should the evacuation area extend?
29. What method should be used to stop the flow from a flammable liquid transportation line if a valve cannot be immediately shut off?
30. What are the two primary problems of oil tank fires?
31. What is the first thing the officer-in-charge should do when arriving at a fire in an oil tank that is located in a refinery?
32. What is generally the most effective extinguishing agent for oil fires?
33. If two foam applicators are available, how should they be set up to extinguish an oil tank fire?
34. How might foam be applied to an oil tank fire if foam applicators are not available?
35. Theoretically, what affect does playing water on the side of a burning tank of oil while applying foam have?
36. What is one of the greatest dangers to firefighters at an oil tank fire?
37. What are the basic requirements for an oil to be capable of boiling over?
38. What oil has all the requirements for a boilover?
39. How long after a fire starts in a crude oil tank should a boilover be anticipated?
40. What is the best action that can be taken when a boilover is expected?

FLAMMABLE GAS EMERGENCIES

1. What are the two flammable gases most commonly encountered by fire departments in emergency operations?
2. What are the two most commonly encountered LPGs?
3. What is meant by a fuel gas?
4. What is one of the primary differences between natural gas and LPG?
5. What is the definition of vapor density?
6. What can be expected of gases having a vapor density of less than 1.0 when they escape from their container?
7. What can be expected of an LPG when it escapes from its container?
8. What are the two basic types of emergencies involving flammable gases to which firefighters respond?
9. What is the primary objective when emergencies involving leaking gas are encountered?
10. What are the three most common methods used to keep an escaping gas from coming into contact with an ignition source?
11. How far beyond the vapor cloud formed by escaping LPG can an ignitible mixture be expected?
12. Where will the primary danger area for potential ignition of an LPG generally be found?
13. How much gas should be expected for every cubic foot of liquid that escapes from an LPG tank?
14. What is a basic principle of fighting fires in combustible gases?
15. How can small natural gas fires in buildings normally be extinguished?
16. What action should be taken when a natural gas fire occurs at the meter?
17. What is the general practice when a fire occurs in the street as a result of a break in a natural gas transmission line?
18. How can minor fires at the relief valve of small LPG storage tanks be extinguished?
19. What is the initial action required when arriving at a large fire involving a LPG tank truck?
20. What type of hose streams are generally required to keep the tank cool and protect the exposures when a large fire involves a LPG tank truck?
21. What is a BLEVE?
22. What generally causes a BLEVE?
23. How far might the danger area for the radiated heat from the fire ball and concussion from the explosion from a BLEVE extend?
24. When do most BLEVEs occur in regard to the tank's contents?
25. What do the records indicate regarding the time between the initiation of the fire and the BLEVE?
26. What are two positive warning signs that should not be ignored regarding a BLEVE?
27. Providing there is a sufficient amount available, what is the best agent for extinguishing a LPG fire?
28. What are the two methods of using dry chemicals on a LPG fire?
29. If it becomes necessary to evacuate an area and let a LPG tanker burn, what is generally the minimum distance from the fire that full evacuation should take place?

ELECTRICAL FIRES

1. What is considered an electrical fire?
2. What is the primary problem for firefighters when fighting electrical fires?
3. What types of portable extinguishers are considered satisfactory for electrical fires?
4. Which type of portable extinguisher that is satisfactory for electrical fires presents a potential corrosion problem?
5. Which extinguishers should be used on sensitive electrical equipment fires?
6. What precautionary measure should be taken when responding to a reported fire in underground electrical equipment?
7. What is the general practice when responding to fires in underground vaults or manholes?
8. What is the first precaution against using water on electrical fires?
9. What is the second precaution?
10. What is the value of a firefighter's boots as an insulator against electrical shocks?
11. How close can an approach to a low-voltage equipment fire be made when using water in the form of fog?
12. What is low voltage?
13. What does past practice indicate as a relatively safe distance for using water on high-voltage equipment?

BRUSH FIRES

1. When is the critical period for a brush fire?
2. What is relative humidity?
3. What are some of the common sources of ignition for brush fires?
4. What is the first and most important factor to be considered in brush firefighting?
5. What is the dual adverse affect that wind has on a brush fire?
6. What is indicated by a brush fire that suddenly stands straight up?
7. What can a canyon be compared to?
8. Where is generally the safest place to work on a brush fire?
9. Where is the most likely place to stop a brush fire?
10. Which ridge is safest to go in on?
11. What dual purposes are served by firebreaks?
12. What are the two types of firebreaks?
13. Under average conditions, how wide should an emergency firebreak be cut?
14. What is probably the most useful universal tool for a brush fire?
15. What are some of the things a shovel can be used for?
16. What are some of the cutting tools used for making an emergency firebreak?
17. What are some of the things for which a helicopter can be used on a brush fire?
18. What is the composition for the most effective helicopter team?
19. How should a house that is considered to be an exposure be prepared to protect it against an advancing brush fire?
20. What is the policy for trying to stop a brush fire advancing up a hill by the use of water?

21. What is the basic principle regarding going down through unburned brush to meet a fire burning uphill?
22. Under what brush conditions may it be possible to make a direct attack on the fire?
23. How much equipment is generally needed to handle a small brush fire?
24. What method of attack is normally used to extinguish a large brush fire?
25. What is meant by a direct attack?
26. What is meant by an indirect attack?
27. What are some of the instructions that should be given by the first-in officer to the responding units at a large brush fire?
28. What position should the first-in officer's crew take?
29. What position should the second-in officer's crew take?
30. What is the objective when it is necessary to use hand tools to make the direct attack?
31. How should dirt be shoveled onto the fire?
32. What is a spot fire?
33. What two conditions must exist for a brush fire to be considered under control?
34. What is meant by cold trailing?
35. How wide should the break be made when cold trailing?
36. What is meant by patrolling?
37. What conditions may necessitate backfiring?
38. Where should a backfire be commenced?
39. What are some of the personnel considerations for use at a brush fire?
40. What are some of the apparatus considerations for use at a brush fire?

PIER AND WHARF FIRES

1. What is a dock?
2. What is a pier?
3. What is a wharf?
4. On what are wharfs built?
5. What is the typical wharf construction in the Los Angeles harbor?
6. How can access to the underside of a wharf be made?
7. What is the policy regarding driving apparatus out onto a wharf if a fire is under the wharf?
8. What action should be taken when the first arriving company is met by a small amount of smoke rising from under a wharf?
9. What action should be taken if a large amount of smoke is issuing from under the wharf?
10. What is the overall strategy for a wharf fire?
11. What tactics are used for setting up a water curtain if the department has a well-trained scuba team?
12. What tactics are used for setting up a water curtain if the department does not have a scuba team?
13. What is the value of the manholes or other openings permanently installed in some wharfs?

14. How is the first direct attack on a wharf fire generally made?
15. How can scuba team members be used in a direct attack?
16. What is the primary task of land-based companies at a wharf fire?

Index